U.S. Army Operator's Manual for M16, M16A1 Rifle

Department of Defense

Pentagon
Publishing

This copy is a reprint which includes current pages from Changes 1 and 2.

TM 9-1005-249-10

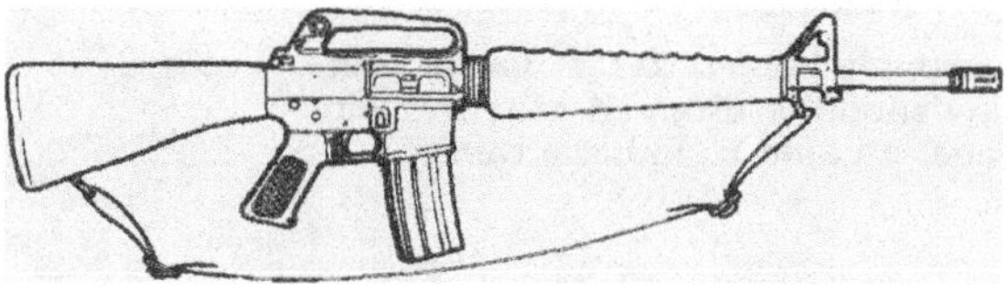

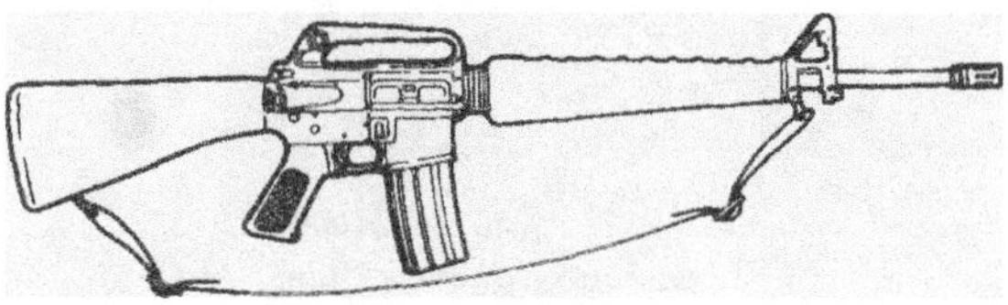

OPERATOR'S MANUAL FOR

RIFLE, 5.56-MM, M16 (1005-00-856-6885)

RIFLE, 5.56-MM, M16A1 (1005-00-073-9421)

HEADQUARTERS, DEPARTMENT OF THE ARMY
FEBRUARY 1985

TM 9-1005-249-10
C2

CHANGE

NO. 2

HEADQUARTERS
DEPARTMENT OF THE ARMY

Washington, DC 11 May 1990

Operator's Manual
for
RIFLE, 5.56-MM, M16
(1005-00-856-6885)
RIFLE, 5.56-MM, M16A1
(1005-00-073-9421)

TM 9-1005-249-10, 11 February 1985, is changed as follows:

Page 2-24. Add the following WARNING before the previous WARNING.

WARNING

With the bolt carrier assembly locked to the rear or in its forward position, if the weapon is dropped or jarred with a loaded magazine in place, it could chamber a round.

By Order of the Secretary of the Army:

CARL E. VUONO
General, United States Army
Chief of Staff

Offlclal:

WILLIAM J. MEEHAN II
Brlgadier General, United States Army
The Adjutant General

Distribution:

To be distributed in accordance with DA Form 12-40, block 135, Operator Maintenance Requirements for Rifle, 5.56 MM, M16, M116A1.

CHANGE

No. 1

HEADQUARTERS
DEPARTMENT OF THE ARMY
Washington, D.C., 9 December 1987

Operator's Manual
for
RIFLE, 5.56-MM, M16
(1005-00-856-6885)
RIFLE, 5.56-MM, M16A1
(1005-00-073-9421)

TM 9-1005-249-10, 11 February 1985, is changed as follows:

Front cover. Add the following under TM 9-1005-249-10: AIR FORCE TO 11W3-5-5-1 COAST GUARD COMDTINST M8370.8

Add, above "FEBRUARY 1985.": "DEPARTMENT OF THE AIR FORCE
COMMANDANTS, COAST GUARD

Page a. The telephone numbers for the AMCCOM Radiological Protection Officer (RPO) are superseded as follows: AUTOVON 793-2964/2965/2966; Commercial (309) 782-2964/2965/2966

Page i.

Add the following to Reporting Errors Block: "Air Force users submit AFTO Form 22, Technical Order System Publications Improvement Report and reply to: WR-ALC/MMEDT, Robins AFB, GA 31098-5000. Coast Guard users submit Publications Correction/Change Report Form to: Commandant, US Coast Guard (G-ODO), Washington, DC 20593-3001."

Supersedure notice is changed as follows:
"*This manual supersedes TM 9-1005-249-10, 1 April 1977, TM 9-1005-249-10-HR, February 1979, and AIR FORCE TO 11W3-5-5-1, 1 August 1966, including all changes."

Page 1-2. Add to end of paragraph: "Air Force users submit Material Deficiency Report (MDR) to: DIR MAT MGT, Robins AFB, GA // MMIRFT // and Quality Deficiency Report to: DIR MAT MGT, Robins AFB, GA // QAY //."

Page 1-3, paragraph 1-4. Add: d. Receiver is made of light-weight aluminum alloys, however, the safety, durability, and function of the rifles are in no way reduced and the portability and logistical values greatly increased, particularly when air transport is used.

Page 1-4. Add: Paragraph 1-7. DESTRUCTION OF ARMY MATERIEL TO PREVENT ENEMY USE. Only your commanding officer can give the order and method to destroy materiel to prevent enemy use. Refer to TM 750-244-7.

Page 1-5.

Change step 5 to read: "Squeezing the trigger releases the hammer, which strikes the firing pin, causing it to impact the primer of the round."

Change step 9 to read: "When round reaches approximate end of barrel, expanding gases from burning propellant pass through gas port, gas tube, and into the bolt carrier assembly, forcing it to the rear. This causes the bolt to extract and eject the spent cartridge case."

Add step 10: "The action spring and buffer assembly returns the bolt carrier assembly forward, stripping a round from the magazine and chambering it."

Page 2-1.

Item (7): Delete " . . . and provides storage for basic cleaning materials."

Item (9): Add: "and provides storage for basic cleaning materials" to end of sentence.

Page 2-6.

Item 2: Under "Equipment is NOT READY/AVAILABLE IF," Add: "There are obstructions in the bore."

Item 3: Under "Équipment is NOT READY/AVAILABLE IF," Add: "Hammer falls."

Page 2-7.

Change the fourth sentence from "Release the trigger." to "Slowly release the trigger."

Add the following NOTE: For the purpose of this test, "SLOW" is defined as 1/4 to 1/2 the normal rate of trigger release.

"Equipment is NOT READY/AVAILABLE IF:" add "hammer fails to fall on the second trigger squeeze or if hammer falls on release of trigger."

Page 2-8. Under "Equipment is NOT READY/AVAILABLE IF:" add "or hammer falls on the second trigger squeeze."

Page 2-9, Item 4. Under "Equipment is NOT READY/AVAILABLE IF:" add "Sights are damaged, missing, or cannot be adjusted."

Page 2-10, Item 5. Under "Equipment is NOT READY/AVAILABLE IF:" add "Magazine catch will not retain or release the magazine."

Page 2-12, Item 8. Add to second paragraph of the WARNING: " . . . or at the end of the day."

Page 2-24.

Change step (1) to read "Pull charging handle assembly (1) rearward, lock bolt, and return charging handle to full forward position. Place selector lever (2) on SAFE."

Change step (2) to read: "Check to see that chamber is clear."

Page 2-31. Paragraph 2-17 is changed as follows:

1 If your rifle fails after performing IMMEDIATE ACTION, perform the following procedures.

2 Clear your rifle (p 2-33), leaving the bolt carrier locked to the rear with the selector lever on SAFE.

3 Check for jammed cartridge case.

4 If a cartridge case is in the chamber, tap out with a cleaning rod.

5 Insert LOADED cartridge magazine (p 2-14) and resume normal operation.

6 If your rifle still fails to fire, clear your rifle (p 2-33) and refer to troubleshooting (p 3-0).

Page 2-32.

Change step 1 to read: "Clear rifle by placing selector lever on SAFE, remove magazine, lock bolt to rear, and check receiver and chamber to ensure that no ammunition is present (p 2-33 and 2-34).

After step 1, add the following WARNING: "Ensure the rifle is clear and that no ammunition is present."

Change step 2 to read: "Visually inspect from the muzzle end and/or insert a cleaning rod into bore to ensure there is not a projectile lodged in bore."

Page 2-40. Add a NOTE before paragraph 2-21: "Protective cap is for field use only. DO NOT store weapon with the protective cap on the barrel. Storing the weapon with the protective cap in place may cause moisture to collect in the barrel."

Page 2-43.

Add a new step (6) to read: "When moving a cold rifle into a warm place, condensation (moisture) will form in and on your weapon. If possible, leave your weapon in a protected, but cold, area outside. When the weapon is brought inside a warm place, it should be disassembled and wiped dry several times as it reaches room temperature."

Renumber steps "(6)" and "(7)" to steps "(7)" and "(8)."

Page 3-0.

Add the following information to paragraph 3-1, a: "Wherever the term CLP or the words lube or lubricant are cited in this TM, it is to be interpreted to mean that CLP, LSA, or LAW can be utilized as applicable. The following constraints must be adhered to (1) Under all but the coldest arctic conditions, LSA or CLP are the lubricants to use on your weapon. Either may be used at -10°F or above. However, do not use both on the same weapon at the same time. (2) LAW is the lubricant to use during cold arctic conditions, +10°F and below. (3) Any of the lubricants may be used from -10°F to +10°F. (4) Do not mix lubricants on the same weapon. The weapon must be thoroughly cleaned during change from one lubricant to another. Dry cleaning solvent (SD) is recommended for cleaning during change from one lubricant to another."

Add a.1 to read: "*Rifle Bore Cleaner.* RBC (item 2.1, app D) may be used to remove carbon buildup in the bore and other parts of the weapon."

Add a.2 to read: "*Dry Cleaning Solvent.* SD (item 3.1, app D) may be used to clean your weapon. Do <u>not</u> use SD on rubber, plastic, or sealed buffers.

Page 3-14. Add a second paragraph to the WARNING: "DO NOT exchange or switch bolt assemblies from one M16/M16A1 to another. It could cause injury to you and damage the rifle."

Page 3-15. Add a CAUTION preceding step 1: "Do not use a screwdriver or any other tool when removing the hand guards, doing so may damage the hand guard and/or slip ring."

Page 3-23. Add a NOTE preceding step 2: Before removing the buffer, assure that the hammer is cocked and the selector lever is NOT set on AUTO."

Page 3-24. Add before the NOTE: "WARNING: Turn magazine away from face, spring is under tension."

Page 3-29. Add the following NOTE after step 2: "Use chamber brush for chamber only."

Page 3-35. Add to step 2: "Check for broken bolt rings."

Page 3-41. Add the following NOTE to the top of the page. "Before installing the buffer assembly, assure that the hammer is cocked and the selector lever is NOT set on AUTO."

Page 3-47. Add to the end of the first CAUTION: ", to prevent damage to automatic sear."

Page 4-1.

Add: "4-1." to AUTHORIZED AMMUNITION.

Add a new paragraph 4-2. CARE, HANDLING, AND PRESERVATION.

4-2. CARE, HANDLING, AND PRESERVATION

a. Protect ammunition from mud, sand, and water. If the ammunition gets wet or dirty, wipe it off at once with a clean, dry cloth. Wipe off light corrosion as soon as it is discovered. Turn in heavily corroded cartridges.

b. Do not expose ammunition to the direct rays of the sun. If the powder is hot, excessive pressure may develop when the rifle is fired.

c. Do not oil or grease ammunition. Dust and other abrasives that collect on greasy ammunition may cause damage to the operating parts of the rifle. Oiled cartridges produce excessive chamber pressure.

Page A-O.

Add to paragraph A-4.: Procedures for Destruction of Equipment to Prevent Enemy Use . . . TM 750-244-7.

Add a new paragraph A-5:

A-5. FORMS

Recommended Changes to Publications and Blank Forms.........DA FORM 2028
Quality Deficiency Report ....SF 368
Equipment Inspection and Maintenance Worksheet ....DA FORM 2404
Technical Order System Publications Improvement Report and Reply ....AFTO FORM 22

Page C-3. Delete item 3, CARTRIDGE CASE DEFLECTOR.

Page C-4. Delete item 2, KIT, ADAPTER SLING.

Page D-3

Change as follows:

Add a new Item 0.1:

0.1	C	1005-00-242-5687	BOTTLE, ASSEMBLY CYLINDRICAL (19204) 8448444	EA

Add a new Item 2.1:

2.1			CLEANING COMPOUND, RIFLE BORE: small arms bore cleaning solution (RBC)	
	C	6850-00-224-6656	2-oz (59.15-ml) bottle	OZ
	O	6850-00-224-6657	8-oz (236.59-ml) can	OZ
	O	6850-00-224-6663	1-gal. (3.79-l) can (81349) MIL-C-372	GL

Page D-4

Change as follows:

Add a new Item 3.1:

3.1	O	6850-00-281-1985	DRY CLEANING SOLVENT (SD) 1-gal. can (58536) A-A-711	GL

Add new Items 3.2 and 3.3:

3.2	O	9150-00-292-9689	LUBRICATING OIL, WEAPONS: (LAW) 1-qt (0.95-l) can (81349) MIL-L-14107	QT
			LUBRICATING OIL, WEAPONS: (LSA), semifluid	
3.3	C	9150-00-935-6597	2-oz (59.15-ml) plastic bottle	OZ
	C	9150-00-889-3522	4-oz (118.30-ml) bottle	OZ
	O	9150-00-687-4241	1-qt (0.95-l) can	QT
	O	9150-00-753-4686	1-gal. (3.79-l) can (81349) MIL-L-46000	GL

By Order of the Secretary of the Army:

CARL E. VUONO
General, United States Army
Chief of Staff

Official:

R. L. DILWORTH
Brigadier General, United States Army
The Adjutant General

Distribution:

To be distributed in accordance with DA Form 12-40, Operator Maintenance requirements for Rifle, 5.56MM, M16, M16A1.

WARNING

All personnel that operate and/or maintain fire control equipment must be aware of the following special precautions.

RADIATION HAZARD
WARNING

TRITIUM (H 3)

Rules and Regulations

Copies of the following rules and regulations are maintained at HQ, AMCCOM, Rock Island, IL 61299-6000. Copies may be requested or information obtained by contacting the AMCCOM Radiological Protection Officer (RPO), AUTOVON 793-3482, Commercial (309) 794-3483.

10CFR Part 19- Notices, Instructions and Reports to Workers; Inspections

10CFR Part 20- Standards for Protection Against Radiation

NRC license, license conditions, and license application

a

WARNING (CONT)

Safety Precautions

The radioactive material used in these instruments is tritium gas (H3) sealed in pyrex tubes. It poses no significant hazard to the repairman when intact. These sources illummate the instrumentation for night operations. Tampering with or removal of the sources in the field is prohibited by Federal law. In the event there is no illumination, notify the local Radlologlcal Protection Officer. Do not attempt to repair or replace the instrument in the field! If skin contact is made with any area contaminated with tritium, immediately wash with nonabrasive soap and water.

Identification

Radioactlve self-luminous sources are identified by means of radioactive warning labels (as above) These labels should not be defaced or removed, and should be replaced immediately when necessary. Refer to the local RPO or the AMCCOM RPO for instructions on handling, storage, or disposal.

Storage and Shipping

All radioactively illuminated instruments or modules which are defective will be evacuated to a depot maintenance activity. These items must be placed in a plastic bag and packaged in the shipping container from which the replacement was taken before evacuation to a higher echelon is made. Spare equipment must be stored in the shipping container as received until installed on the weapon. Storage of these items is recommended to be in an outdoor shed type storage or unoccupied building.

FIRST AID

For further information on first aid, see FM 21-11

b

WARNING (CONT)

To avoid accidental firing, BE SURE WEAPON IS CLEAR. Failure to do so could result in serious injury or death.

Be sure the cam pin is installed in the bolt group. If it isn't, your rifle can still fire and will explode causing injury or death.

If you're using the blank firing attachment, don't use any other ammmunition except the blank round, M200.

Do NOT exchange or switch bolt assemblies from one M16/M16A1 to another. It could cause damage to both you and the rifle.

DON'T OVERHEAT M16/M16A1 RIFLE BARRELS. Sustained firing of the M16/M16A1 Rifle Will rapidly raise the temperature of the barrel to a critical point.

Firing 140 rounds, rapidly and continuously, will raise the temperature of the barrel to the COOKOFF POINT. At this temperature, any live round remaining in the chamber for any reason may cook off (detonate) in as short a period as 10 seconds.

If the cookoff point (or temperature) is felt possible, weapon should be immediately cleared and allowed to cool.

Sustained rate of fire for the M16/M16A1 Rifle is 12-15 rounds per minute. This is the actual rate of fire that a weapon can continue to deliver for an indefinite length of time without serious overheating.

Sustained rate of fire should never be exceeded except under circumstances of extreme urgency

If your bolt fails to unlock and you try to free it by banging the buttstock on the ground, keep clear of the muzzle.

WARNING (CONT)

If there's water in the barrel, don't fire the rifle, It could explode,

If you experience a noticeable difference in sound or recoil, STOP FIRING. Either condition could indicate an Incomplete propellant burn and a bullet still in the bore. Retract bolt slowly and remove fired cartridge case. Clear weapon and check for unburned powder grams in the receiver or bore and for a bullet in the bore. Remove unburned propellant or bullet from bore before resuming firing or barrel could explode. If bullet is lodged in bore, turn in rifle to the unit armorer,

If rifle stops firing with a live round in the chamber of a hot barrel, remove the round fast. However, during training, if you cannot remove it within 10 seconds, wait 15 minutes with the rifle pointing in a safe direction. This way you won't get hurt by a possible ammunition cookoff, which could happen 10 seconds after contact with a hot chamber. Clear rifle.

Use only authorized ammunition that is manufactured to US specifications,

Blank ammunition should not be fired toward personnel within 20 feet or less from the muzzle, because fragments of a closure wad or particles of unburned propellant might inflict injury within that range.

If you go by all the instructions in this book, and perform preventive maintenance (PM), your M16/M16A1 rifle will operate properly. If you've done your part and it fails to perform properly, turn in your M16/M16AI rifle to your unit armorer.

TECHNICAL MANUAL*
No. 9-1005-249-01

HEADQUARTERS
DEPARTMENT OF THE ARMY
Washington, DC *11 February 1985*

Operator's Manual
for
RIFLE, 5.56-MM, M16
(1005-00-856-6885)
RIFLE, 5.56-MM, M16A1
(1005-00-073-9421)

REPORTING ERRORS AND RECOMMENDING IMPROVEMENTS

You can help improve this manual. If you find any mistakes or if you know of a way to improve the procedures, please let us know. Mail your letter or DA Form 2028 (Recommended Changes to Publications and Blank Forms) direct to: Commander, US Army Armament, Munitions and Chemical Command, ATTN: AMSMC-MAS, Rock Island, IL 61299-6000. A reply will be furnished to you.

*This manual supersedes TM 9-1005-249-10, 1 April 1977, including all changes.

TABLE OF CONTENTS

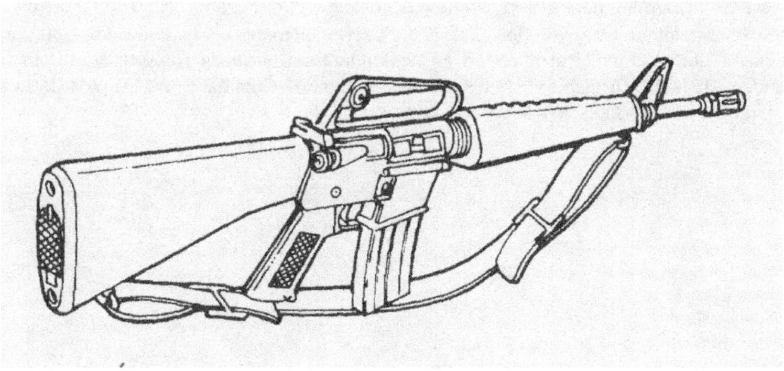

M16A1 RIFLE

CHAPTER 1
INTRODUCTION

Section I. GENERAL INFORMATION

1-1. SCOPE.

a. Type of Manual: Operator's

b. Model *Number and Equipment Name:* M16/M16A1 5.56-mm Rifle.

c. *Purpose of Equipment:* To provide personnel an offensive/defensive capability to engage targets in the field.

1-2. MAINTENANCE FORMS AND RECORDS. Department of the Army forms and procedures used for equipment maintenance will be those prescribed by DA PAM 738-750, The Army Maintenance Management System (TAMMS).

1-3. REPORTING EQUIPMENT IMPROVEMENT RECOMMENDATIONS (EIR's). Ifyourrifle needs improvement,let usknow. Sendusan EIR. You, theuser, arethe only one who can tell us what you don't like about your equipment. Let us know why you don't like the design or performance. Put it on an SF 368 (Quality Deficiency Report). Mail it to us at Commander, US Army Armament, Munitions and Chemical Command, ATTN: AMSMC-QAD, Rock Island, IL 61299-6000. We'll send you a reply.

Section II. EQUIPMENT DESCRIPTION

1-4. EQUIPMENT CHARACTERISTICS, CAPABILITIES, AND FEATURES.

a. The M16/M16A1 rifle is lightweight, air cooled, gas operated, magazine fed, and shoulder fired.

b. The rifle may be fired with selector lever in the automatic or semiautomatic position.

c. It also provides personnel an offensive/defensive capability to engage targets in the field.

1-5. DIFFERENCES BETWEEN MODELS. The 5.56-mm Rifle M16 does not contain the forward assist assembly contained on the 5.56-mm Rifle MI 6A1. Both models may be equipped with the low light level sight assembly.

1-6. EQUIPMENT DATA.

Weight:

Rifle M16, without cartridge magazine and sling 6.351b

Rifle M16A1, without cartridge magazine and sling . . . . 6.55 lb

Overall length:

Rifle w/flash suppressor . . . .39 in.

Rifle w/bayonet-knife . . . .44.25 in.

Maximum rate of fire:

Semiautomatic . . . .45/65 rounds/m

Automatic . . . .150/200 rounds/m

Maximum effective range . . . .460 meters

Section III. TECHNICAL PRINCIPLES OF OPERATION

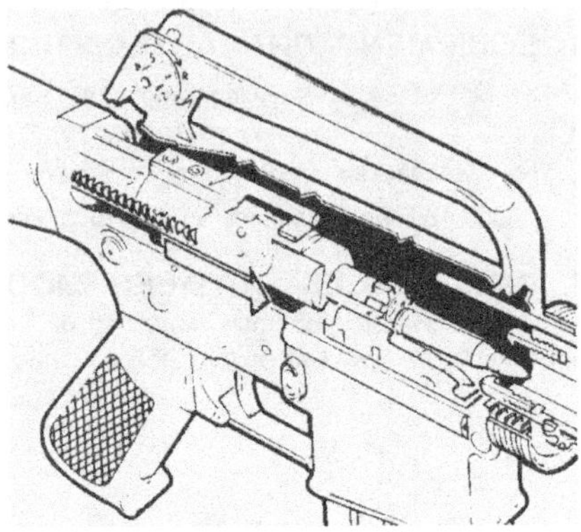

NOTE

Magazine may be loaded with bolt assembly open or closed.

1 Place selector lever on SAFE.
2 Insert loaded caftridge magazine in magazine well and chamber a round.
3 Face the target, move the selector lever from SAFE to SEMI or AUTO, and place the rifle to your shoulder.
4 Aline the front and rear sight with the target and squeeze the trigger.
5 Squeezing the trigger releases the firing pin and allows it to impact the primer on the round.
6 The primer ignites the propellant in the round.
7 Gas from the burning propellant pushes the projectile along the barrel of the rifle.
8 The rifling in the barrel causes the projectile to rotate which provides stability during flight to the target.
9 When round reaches approximate end of barrel, expanding gases from burning propellant pass out through gas port and into gas tube. Gas goes into bolt carrier assembly, ejects old cartridge, and chambers a new round.

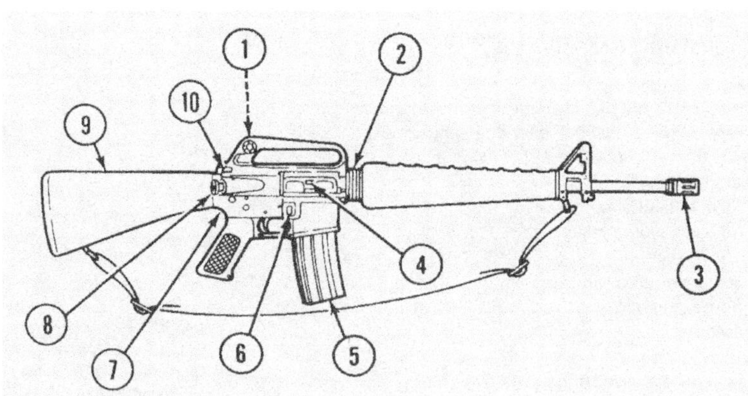

RIGHT SIDE

CHAPTER 2
OPERATING INSTRUCTIONS

Section I. DESCRIPTION AND USE OF OPERATOR'S CONTROLS AND INDICATORS

2-1. M16/M16A1 MECHANICAL CONTROLS.

REAR SIGHT (1) - zeros weapon and engages targers to 460 meters.

HAND GUARD SLIP RING (2) - keeps hand grards in place.

FLASH SUPPRESSOR (3)- reduces the amount of flash from muzzle when weapon is fired.

EJECTION PORT COVER (4) - protects upper receiver from foreign matter when weapon is not in use. Keep port cover closed when not used.

CARTRIDGE MAGAZINE (5) - supplies 30 rounds of ammunition to the weapon.

MAGZINE CATCH BUTTON (6) - releases cartridge magazine (5) from weapon when pushed.

LOWER RECEIVER AND EXTENSION ASSEMBLY (7)- provides firing control for the weapon and provides storage for basic cleaning materials.

FORWARD ASSIST ASSEMBLY (M16A1 ONLY) (8) - ensures that bolt is fully forward and locked.

SHOULDER GUN STOCK ASSEMBLY (9) - stabilizes rifle.

CHARGING HANDLE ASSEMBLY (10) - cocks weapon when preparing to fire or clearing weapon.

2-1. M16/M16A1 MECHANICAL CONTROLS (CONT).

FRONT SIGHT POST (11) - adjustable for elevation

CARRYING HANDLE ASSEMBLY (12) - provides the means for hand-carrying the rifle,

SELECTOR LEVER (13) - arms the rifle in SEMI or AUTO or safes the rifle.

TRIGGER (14) - controls the firing of the weapon

SMALL ARMS SLING (15) - provides the means for shoulder-carrying the weapon,

BOLT CATCH (16) - moves the key and bolt carrier assembly forward when depressed,

BAYONET STUD (17) - holds bayonet in place

UPPER RECEIVER AND BARREL ASSEMBLY (18) - directs the projectile upon firing.

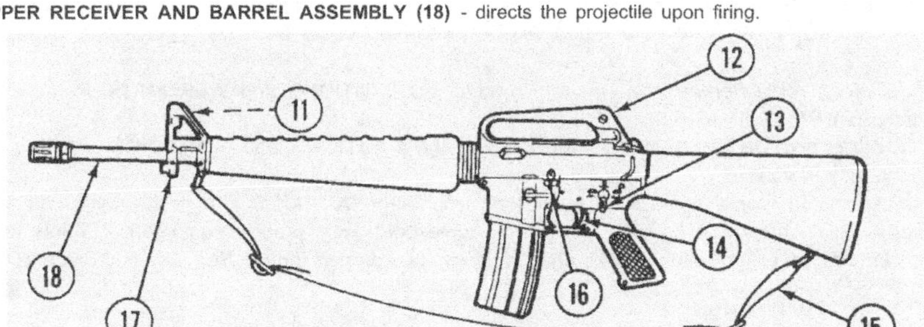

LEFT SIDE

2-2. THROW AWAY THE WHITE GLOVES FOR RIFLE INSPECTIONS.

a. CLP will leave a film, or layer, of Teflon that builds up over a period of time This is one of the benefits of using CLP, especially in combat, where you might not have time to lube your rifle as often as you think it needs it. So, throwaway your white gloves. But carry a rag with you to wipe your harnds when you inspect rifles because your fingers are going to get, a little slick if your troops have used the right amount of CLP on their rifles.

b. Now, if you are going to inspect rifles the following day, give your troops a few minutes to wipe their rifles down again. Remember, CLP is a cleaner and it never stops working. So, overnight while the Teflon has been forming a film for lubrication, the cleaning solvents in the CLP have been at work in the nooks and crannies (actually in the pores of the metal) seeking out carbon and firing residue.

2-3. ARMORY AND ORDNANCE INSPECTORS. You professionals who inspect weapons and armories should be the experts who are out there insisting that a light coat of CLP be kept on rifle metal parts at all times. Since it's always at work cleaning, expect "a little" carbon to be present. If there is doubt as to whether the rifle is or was cleaned properly or not, field-strip it. As the expert, you should be able to tell from plenty of other indicators, e.g., carbon under extractor, on firing pin or bolt, etc. whether or not the rifle was cleaned, lubed, and preserved properly.

Section II. PREVENTIVE MAINTENANCE CHECKS AND SERVICES (PMCS)

2-4. GENERAL. Perform after (A) operations PMCS if: you are the assigned operator and the weapon has been stored in the arms room and not used for a period of 90 days, or you have been issued the weapon for the first time.

NOTE

An inactive weapon is a weapon, whether assigned or not assigned to an individual, that is stored in an arms room for a period of 90 days. Performance of normal cleaning (PMCS) of an inactive weapon will be performed every 90 days,

a. *Before You Operate.* Always keep in mind the CAUTIONS and WARNINGS. Perform your before (B) PMCS.

b. *While You Operate.* Always keep in mind the CAUTIONS and WARNINGS. Perform your during (D) PMCS.

c. *After You Operate.* Be sure to perform your after (A) PMCS.

d. *If Your Equipment Fails to Operate.* Troubleshoot with proper equipment. Report any deficiencies to organizational maintenance using the proper forms. See DA PAM 738-750.

2-5. PMCS PROCEDURES. The PMCS table lists those required checks and services to be performed to ensure accurate performance of the rifle. When recording results of PMCS, entries in the PMCS item No. column shall be used for the TM Item No. column on DA Form 2404. The third column lists the item to be inspected. The fourth column contains conditions that make the rifle not ready/availabie because of inability to perform its primary combat mission. If anything looks wrong, and you cannot correct it yourself, notify organizational maintenance,

OPERATOR PREVENTIVE MAINTENANCE CHECKS AND SERVICES (PMCS)

B - Before Operation D - During Operation A - After Operation

Item No.	Interval B	Interval D	Interval A	ITEM TO BE INSPECTED Procedure	Equipment is NOT READY/AVAILABLE IF:

WARNING

Before starting an inspection, be sure to clear the rifle (p 2-33). Do not squeeze the trigger until the rifle has been cleared. Inspect the chamber to ensure that it is empty and no ammunition is in position to be chambered. Do not keep live ammunition near work area.

Item No.	B	D	A	ITEM TO BE INSPECTED Procedure	Equipment is NOT READY/AVAILABLE IF:
1	●			**ESSENTIAL TOOLS AND EQUIPMENT.** Check the authorized tools and equipment and auxiliary equipment for completeness and serviceability. (See appendix C, section II.)	

OPERATOR PREVENTIVE MAINTENANCE CHECKS AND SERVICES (PMCS) (CONT)

B- Before Operation D- During Operation A After Operation

Item No.	Interval B	Interval D	Interval A	ITEM TO BE INSPECTED Procedure	Equipment is NOT READY/AVAILABLE IF:
2	●			**MAINTENANCE READINESS.** Clear and clean bore with dry swab (Item 5, app D) (p 3-31).	
3				**SELECTOR LEVER FUNCTIONING.**	
	●			a. Cock the rifle and place the selector lever in SAFE position. Squeeze the trigger; the hammer should not fall.	Selector lever does not function.

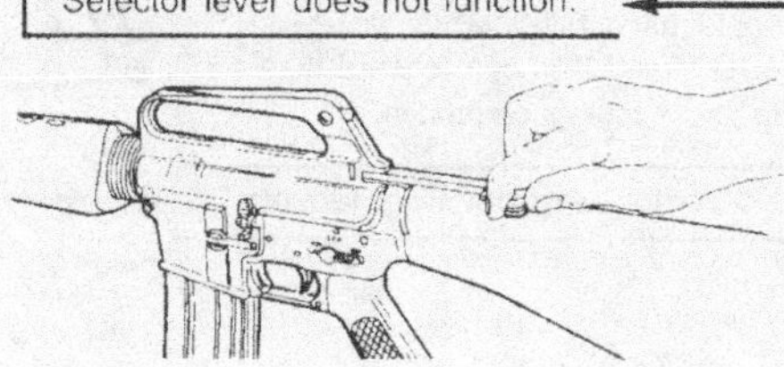

OPERATOR PREVENTIVE MAINTENANCE CHECKS AND SERVICES (PMCS) (CONT)

B - Before Operation D- During Operation A - After Operation

Item No.	Interval B	D	A	ITEM TO BE INSPECTED Procedure	Equipment is NOT READY/AVAILABLE IF:
	●			SELECTOR LEVER FUNCTIONING (CONT) b. SEMI/Posifiorr. Squeeze tigger; hammer should fall. Hold triggerto the rear and recock rifle. Release the trigger. You should hear a click as you release the trigger. Again squeeze trigger; hammer should fall.	Selector lever does not function.

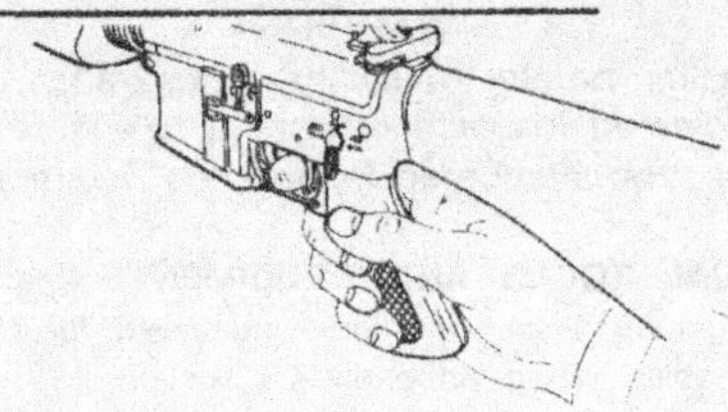

OPERATOR PREVENTIVE MAINTENANCE CHECKS AND SERVICES (PMCS) (CONT)

B- Before Operation D- During Operation A- After Operation

Item No.	Interval B	D	A	ITEM TO BE INSPECTED Procedure	Equipment is NOT READY/AVAILABLE IF:
	●			SELECTOR LEVER FUNCTIONING (CONT). c. *AUTO Position.* Cock the rifle. Squeeze the trigger; hammer should fall Hold trigger to the rear and cock the rifle. Release the pressure on the trigger and squeeze it to the rear again. The hammer should not fall because it should have fallen when the bolt was allowed to move forward during the cocking sequence.	Selector lever does not function.

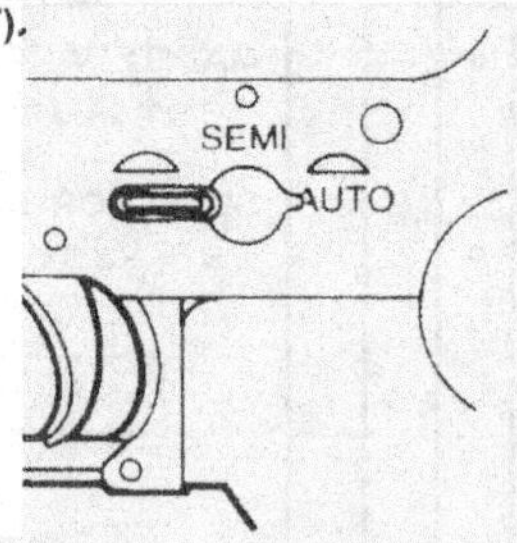

OPERATOR PREVENTIVE MAINTENANCE CHECKS AND SERVICES (PMCS) (CONT)

B - Before Operation D - During Operation A - After Operation

Item No.	Interval B	Interval D	Interval A	ITEM TO BE INSPECTED Procedure	Equipment is NOT READY/AVAILABLE IF:
4	•			**RIFLE SIGHTS (ZERO ADJUSTMENT).** Move front (1) and rear (2) sight to make sure they can be adjusted. Return sights to zero setting of your rifle (p 2-19).	

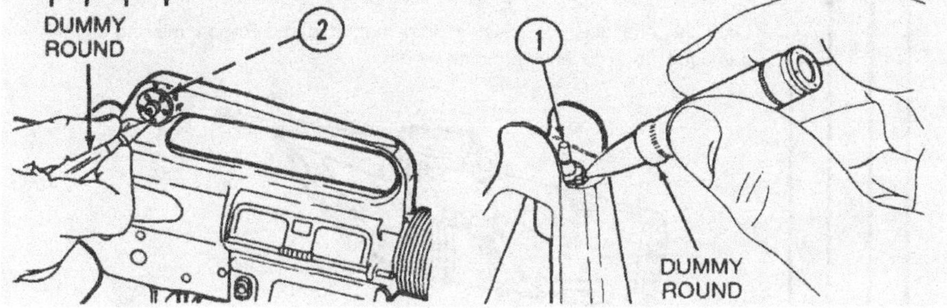

OPERATOR PREVENTIVE MAINTENANCE CHECKS AND SERVICES (PMCS) (CONT)

B - Before Operation D - During Operation A - After Operation

Item No.	Interval B	Interval D	Interval A	ITEM TO BE INSPECTED Procedure	Equipment is NOT READY/AVAILABLE IF:
5	•			**MAGAZINE CATCH (FUNCTION).** Insert magazine (l) into the well, The magazine catch should hold the magazine in place. Pressing the magazine catch button (2) should release the magazine. To adjust the magazine catch, use cleaning rod to press in on the magazine catch button until the left side of the magazine catch (3) sticks out beyond the receiver. To tighten, turn the magazine catch clockwise; to loosen, turn it counterclockwise.	

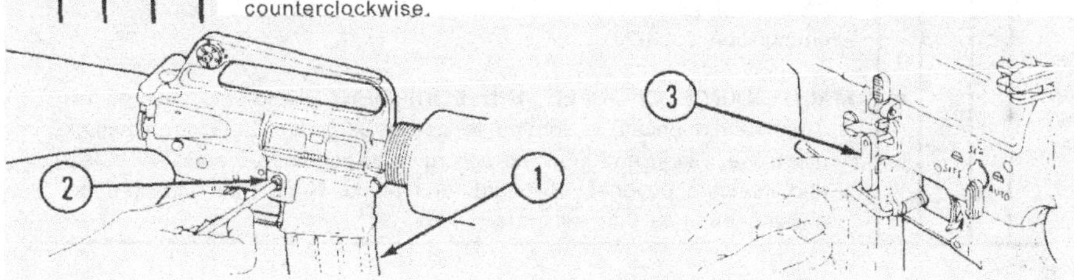

OPERATOR PREVENTIVE MAINTENANCE CHECKS AND SERVICES (PMCS) (CONT)

B - **Before Operation** D - **During Operation** A - After Operation

Item No.	Interval B	Interval D	Interval A	ITEM TO BE INSPECTED Procedure	Equipment is NOT READY/AVAILABLE IF:
6	•			**VISUAL INSPECTION OF RIFLE.** **WARNING** Be sure rifle is clear. Refer to pages 2-33 thru 2-35. Look the rifle over for missing or damaged parts. Report missing or damaged parts to emit armorer.	

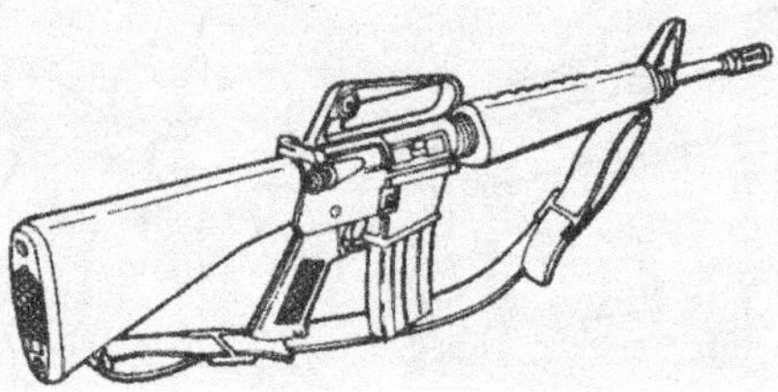

OPERATOR PREVENTIVE MAINTENANCE CHECKS AND SERVICES (PMCS) (CONT)

B - Before Operation D - During Operation A - After Operation

Item No.	Interval B	Interval D	Interval A	ITEM TO BE INSPECTED Procedure	Equipment is NOT READY/AVAILABLE IF:
7		•		**PERIODIC INSPECTION OF RIFLE.** Periodically check rifle to make sure It's clean and there is no foreign material in bore. If foreign material is in bore, clean bore (p 3-29).	
8		•		**MAINTENANCE PERFORMED DURING FIRING OPERATIONS.** **WARNING** Be sure rifle is clear. Refer to pages 2-33 thru 2-35. Clean and lubricate rifle after firing approximately 200 rounds of ammunition (p 3-28).	
9			•	**MAINTENANCE OF RIFLE AND EQUIPMENT.** Field-strip rifle(p3-14). Clean and lubricate according to pages 3-28 thru 3-33. Disassemble magazine. Clean and lubricate according to pages 3-26 and 3-27. Clean and lubricate bayonet, scabbard, and bipod. Report all damaged or missing parts to unit armorer.	

Section III. OPERATION UNDER USUAL CONDITIONS

2-6. PREPARATION FOR STORAGE IN ARMS ROOMS.

WARNING

Be sure rifle is clear. Refer to pages 2-33 thru 2-35

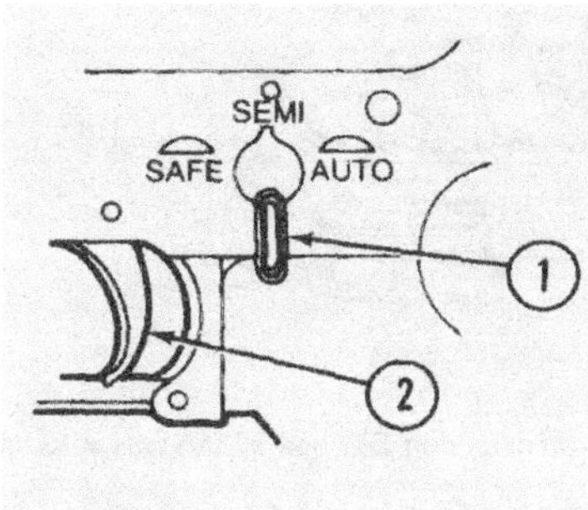

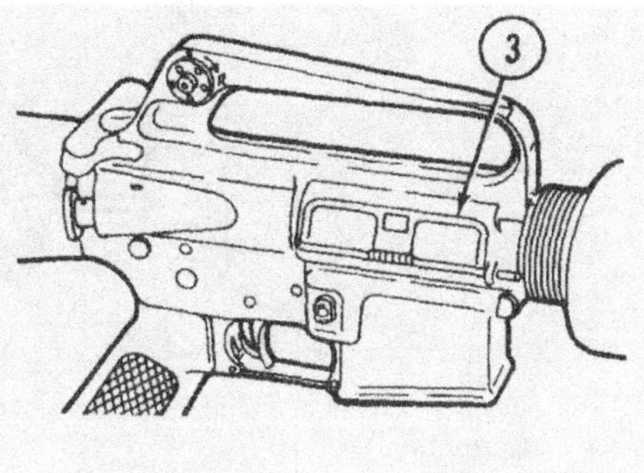

1 Place selector lever (1) on SEMI and squeeze trigger (2) to uncock rifle.

2 Close ejection cover (3) and place in rack.

2-7. ASSEMBLY AND PREPARATION FOR FIRING - CLEAN AND LUBRICATE.

1 Clear your rifle (p 2-33).

2 Look for fouling in bore and chamber.

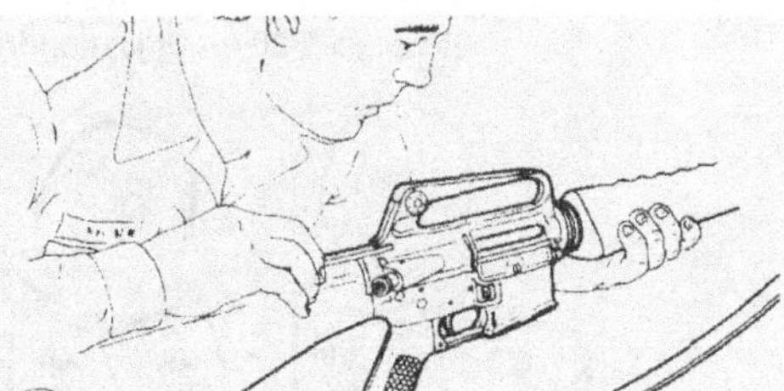

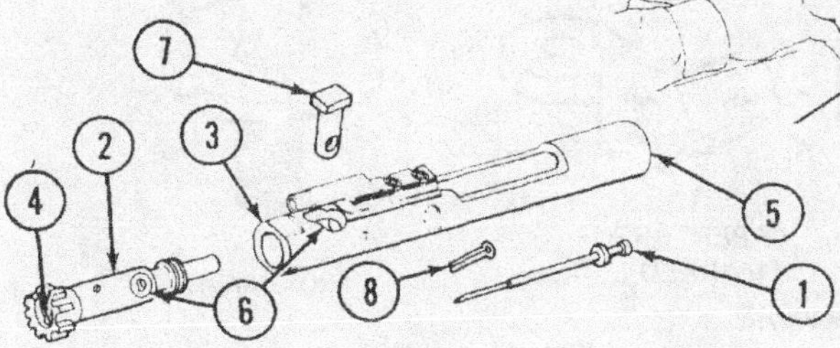

3 Remove and disassemble (p 3-17) bolt carrier assembly and clean carbon and oil from firing pin (1) and all surfaces of bolt assembly (2) and bolt carrier assembly (3) with dry swabs (item 5, app D). Clean firing pine hole (4) and bolt carrier key (5) with pipe cleaner (item 2, app D). Lightly coat with CLP (item 1, app D). Pay special attention to bolt cam pin area (6). Lightly lube bolt cam pin (7) and firing pin retaining pin (8).

CAUTION
Don't bend or flex cleaning rod.

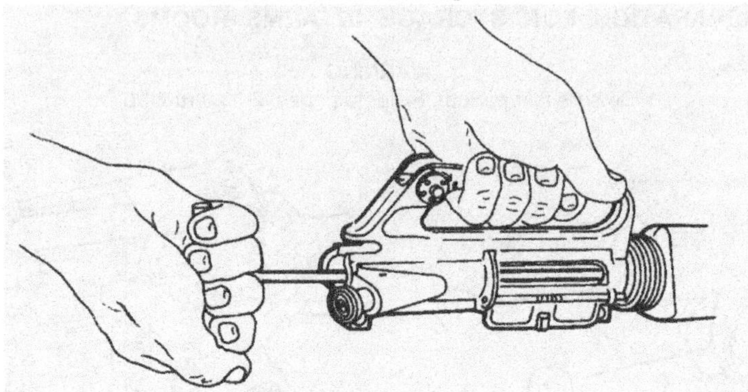

4 "Swab out" (from chamber to muzzle). Make sure swab (item 5, app D) goes all the way through flash suppressor.

5 Clean and lubricate lugs (p 3-29).

6 Reassemble and install bolt carrier assembly (p 3-45).

2-8. INITIAL ADJUSTMENTS - STANDARD DAYLIGHT SIGHT SYSTEM.

NOTE
See page 2-20 for adjusting front and rear sights.

REAR SIGHT

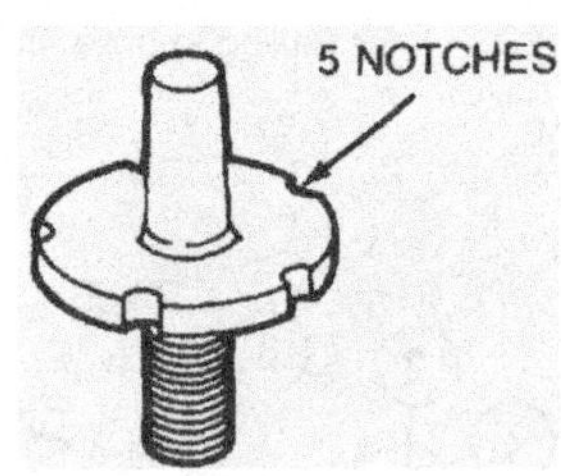

FRONT SIGHT

1 REAR SIGHT - Has two apertures for range.

- **Use the unmarked aperture for targets from 0 -300 meter.**
- **Use the aperture marked L for targets from 300-400 meters.**

2 FRONT SIGHT - Has five notches of elevation per revolution.

2-9. INITIAL ADJUSTMENTS - LOW LIGHT LEVEL SIGHT SYSTEM.

NOTE

See page 2-21 for adjusting the sights.

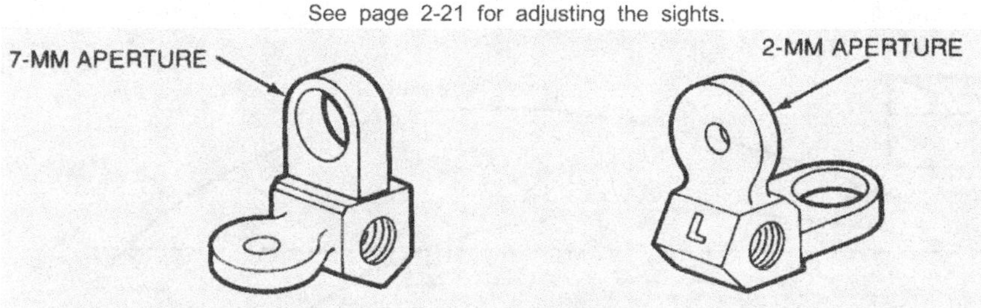

REAR SIGHT

1 Use the low light level sight system when visibility is limited. Not every rifle will have this sight system.

2 REAR SIGHT - Has two apertures.

- Use 7-mm rear sight aperture (unmarked aperture) for night firing and when visibility is limited.
- Use 2-mm rear sight aperture (aperture marked L) to zero the weapon and to hit targets Up to 460 meters under normal conditions.

2-9. INITIAL ADJUSTMENTS - LOW LIGHT LEVEL SIGHT SYSTEM (CONT).

WARNING

RADIATION HAZARD

The front sight post contains a small glass vial of radioactive Tritium H 3. Take care not to bump, abuse, tamper or alter the post in any manner.

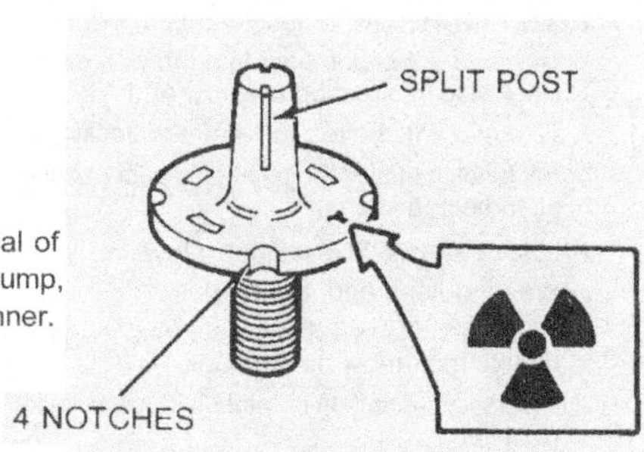

FRONT SIGHT

CAUTION

Do not blacken or soot-up the front sight,

3 FRONT SIGHT - Has only four notches of elevation per revolution, Split post holds vial of luminous material. Material can be seen from two sides only. If you cannot see the vial after zeroing, turn the sight one click down for use during periods of limited visibility.

2-10. INITIAL ADJUSTMENTS - BATTLESIGHT ZERO.

NOTE

To zero the rifle, adjust the front sight (elevation) and the rear sight (windage) so that you can hit aiming point at a given range.

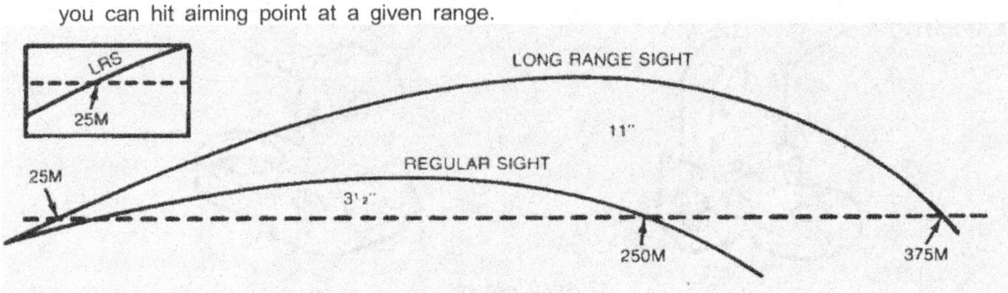

1 Battlesight zero is that setting on the M16's sights which will cause the point of aim and strike of the bullet to be the same at 250 meters.

2 When using the L-marked aperture, the path of the bullet will cross the line of sight at 25 meters. Hence, zeroing is now conducted with point of aim and point of impact being the same.

3 The 25 meter zero target (NSN 6920-01-167-1392) has complete zeroing instructions printed on its face. The target is printed on both sides: one side is printed for the standard sights and the other side is printed for use with the low light level sight system. The grid printed on each target is set up so that one click of elevation of windage is equal to one block change in elevation or windage.

2-11. INITIAL ADJUSTMENTS - SIGHT ADJUSTMENT (LOW LIGHT LEVEL AND STANDARD DAYLIGHT SYSTEMS).

REAR SIGHT To adjust windage, depress detent and rotate drum to direction you want:

1 To move point of impact to right, turn drum clockwise in directlon of arrow and letter R

2 To move left, move drum counterclockwise

3 Each notch moves the point of impact of bullet as Indicated in chart

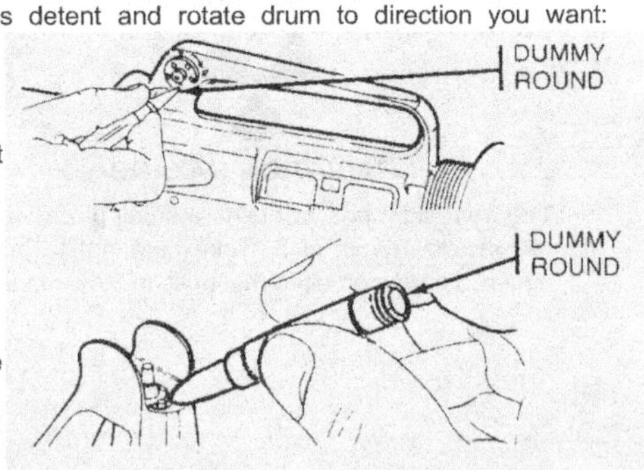

FRONT SIGHT To adjust elevation, depress detent and rotate post:

1 To raise strike of bullet, rotate post in the direction of arrow marked up

2 Reverse the direction of rotation to lower strike of bullet

3 Each notch moves the point of impact of bullet as indicated in chart

IMPACT (1 CLICK)		DISTANCE
STANDARD SYSTEM	LOW LIGHT LEVEL SIGHT SYSTEM	AT
0.7cm (17/64 in.)	0.9cm (23/64 in.)	25 meters
2.8cm (1-3/32 in.)	3.5cm (1-3/4 in.)	100 meters
5.6cm (2-13/64 in.)	7.0cm (2-3/4 in.)	200 meters

2-12. INITIAL ADJUSTMENTS - USING LOW LIGHT LEVEL SIGHT SYSTEM.

DAYLIGHT FIRING

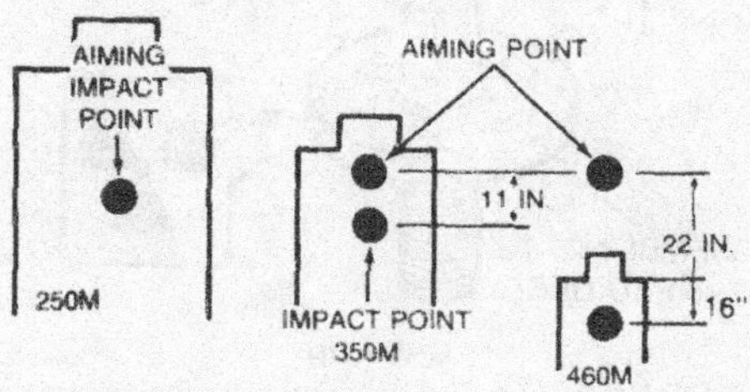

1 Use aperture marked L.

2 Effective range is 250 meters (original battlesight zero); beyond that use hold-off (aiming above desired point of impact).

3 Aim about 11 inches above top of target at 350 meters and about 22 inches above top of target at 460 meters.

4 To become and remain proficient, practice hold-off,

2-12. INITIAL ADJUSTMENTS - USING LOW LIGHT LEVEL SIGHT SYSTEM (CONT).

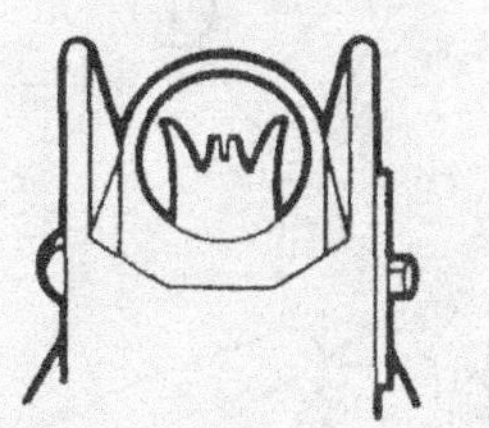

NIGHT AND LIMITED VISIBILITY

5 Use unmarked (7-mm) aperture

6 Use daylight hours procedure to obtain good sight picture.

7 After detecting target, aline sight by centering top of lminous portion of front sight post within 7-mm aperture on target, and fire.

8 Under certain light conditions, you can see front sight post, but you can't determine whether you are looking through, above, or to the side of rear sight aperture.

9 Practice positioning stock against shoulder and looking through rear aperture.

2-13. INITIAL ADJUSTMENTS - CARE AND CLEANING OF FRONT SIGHT.

HANDLING AND CARE

WARNING

RADIATION HAZARD

The low light level front sight post has a small **glass vial containing radioactive Tritium H 3.** Take care not to bump, abuse, alter or tamper with the post in any manner.

CAUTION

Do not blacken or soot-up the front sight.

NOTE

Frequently check the sight post for damage. If damage is evident or suspected, have sight post checked by the unit armorer.

CLEANING OF FRONT SIGHT POST

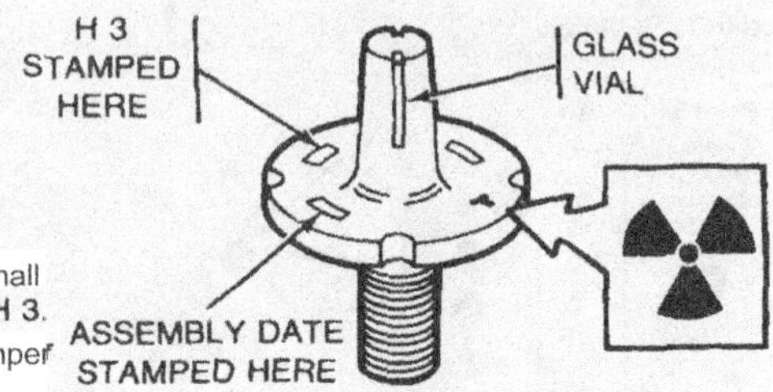

CAUTION

DO NOT use a wire brush to clean the sight.

Clean front sight post with:

- Toothbrush
- Swabs (item 5, app D)
- Cleaner, Lubricant and Preservative (CLP) (item 1, app D).

2-14. OPERATING PROCEDURE - LOADING.

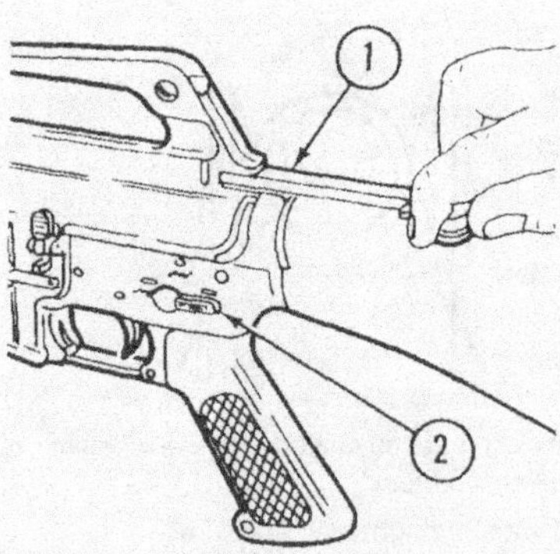

WARNING

Point muzzle in safe direction.

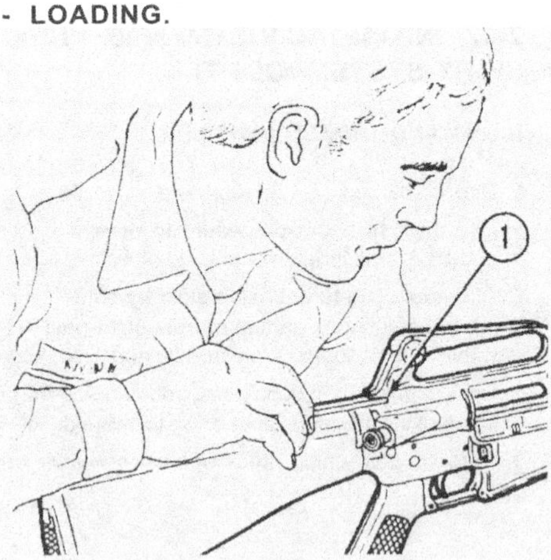

1 Pull charging handle assembly (1) rearward, lock bolt, and release charging handle. Place selector lever (2) on SAFE.

2 Pull charging handle assembly (1) rearward and check to see that chamber is clear, Release charging handle assembly,

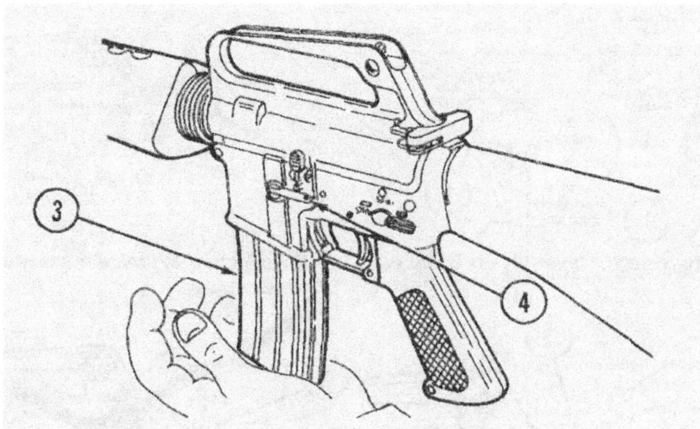

NOTE

Magazine may be loaded with bolt assembly open or closed.

3 Push upward on cartridge magazine (3) until magazine catch (4) engages and holds cartridge magazine.

4 Tap upward to make sure cadridge magazine is seated correctly.

2-15. OPERATING PROCEDURE- CHAMBERING AND FIRING A ROUND.

BOLT ASSEMBLY OPEN

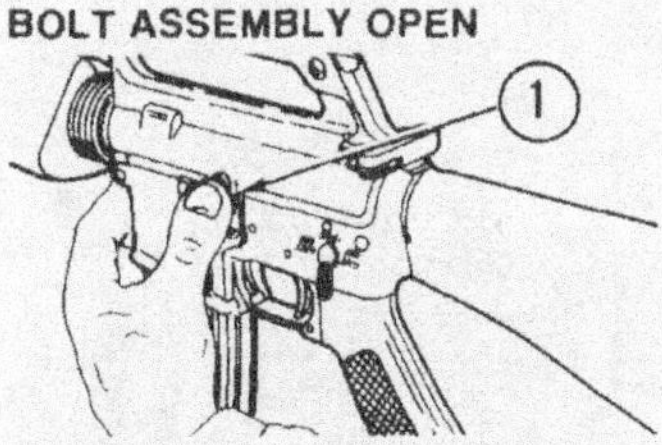

1 Depress upper portion of bolt catch (1) to release the bolt.

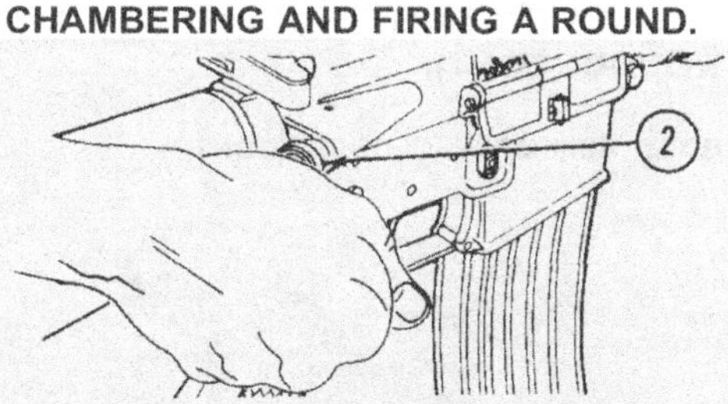

2 Tap forward assist assembly (2) to ensure bolt is fully forward and locked.

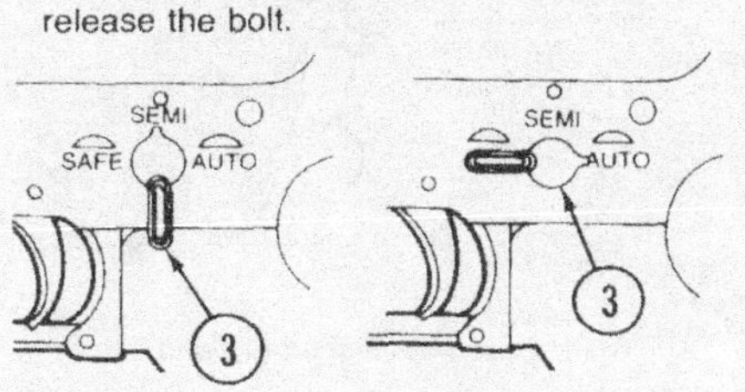

3 Move selector lever (3) to SEMI or AUTO.

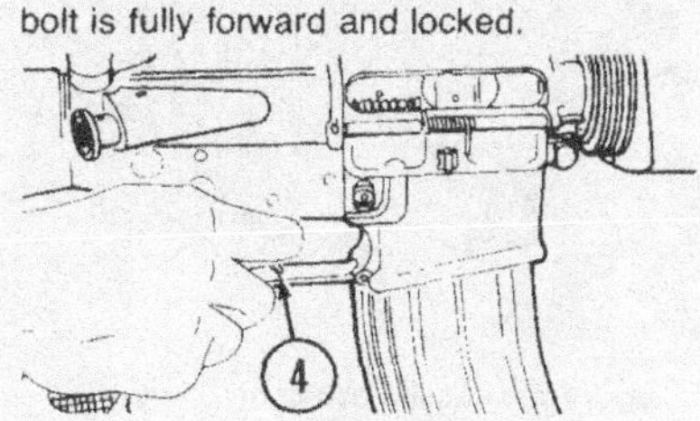

4 Squeeze the trigger (4) and fire.

BOLT ASSEMBLY CLOSED

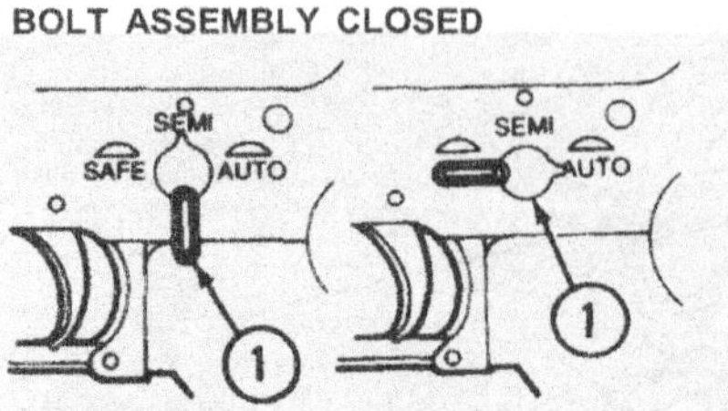

1 Place the selector lever (1) on SEMI or AUTO.

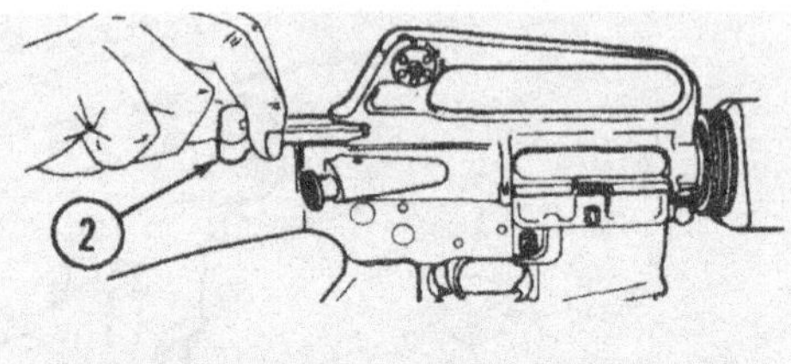

2 Pull charging handle assembly (2) all the way back.

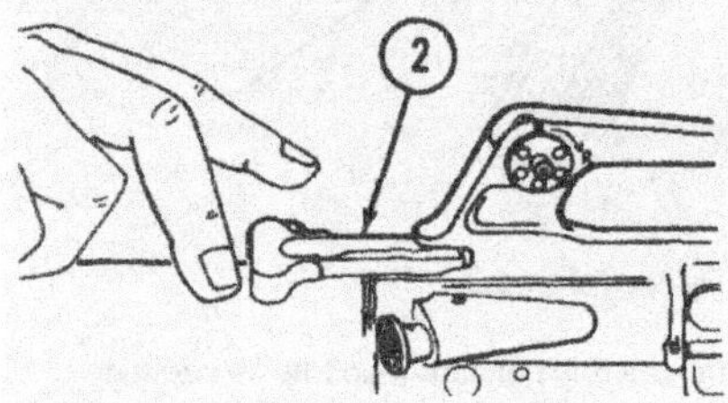

3 Release the charging handle assembly (2).

4 Never "ride" the charging handle assembly (2). Let it go on its own.

2-15. OPERATING PROCEDURE - CHAMBERING AND FIRING A ROUND (CONT).

BOLT ASSEMBLY CLOSED (CONT)

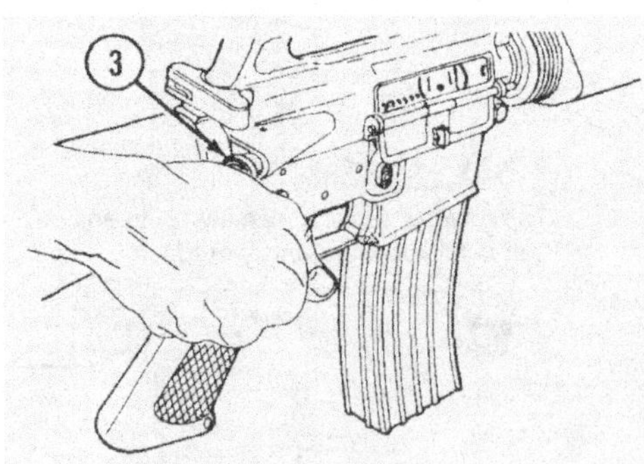

5 Tap forward assist assembly (3) to ensure bolt is fully forward and locked (M16A1 only).

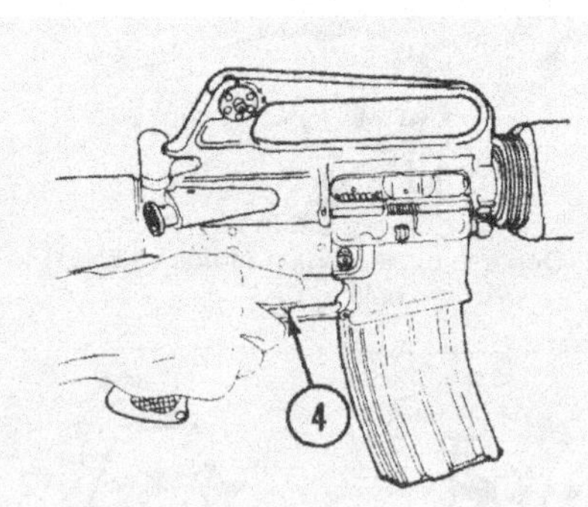

6 Squeeze the trigger (4) and fire.

2-16.OPERATINGPROCEDURE-IMMEDIATE ACTION.

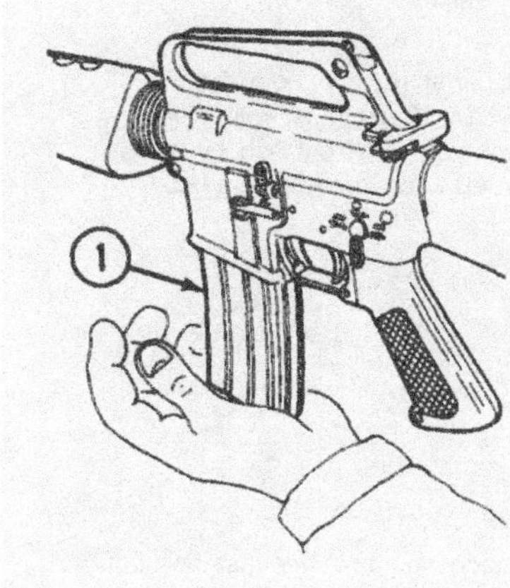

1 Slap upward on cartridge magazine (1) to make sure it's properly seated.

2 Pull charging handle assembly (2) all the way back. Observe ejection of case or cartridge. Inspect chamber (3) and check for obstruction. If chamber is not clear, apply remedial action (p 2-31)

2-16. OPERATING PROCEDURE-IMMEDIATE ACTION (CONT).

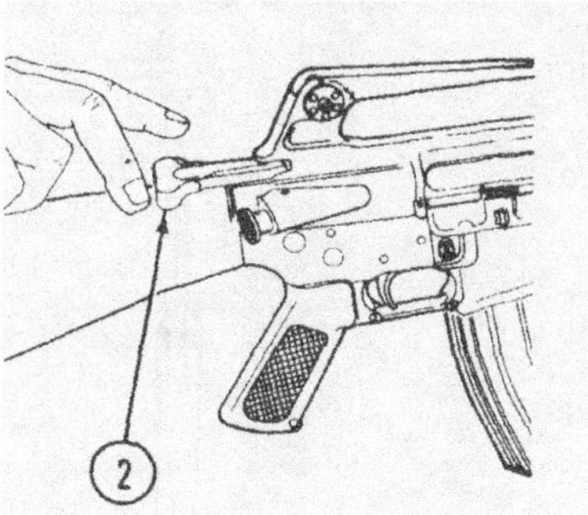

3 Release charging handle assembly (2) to feed new round. (Don't ride the charging handle assembly (2).)

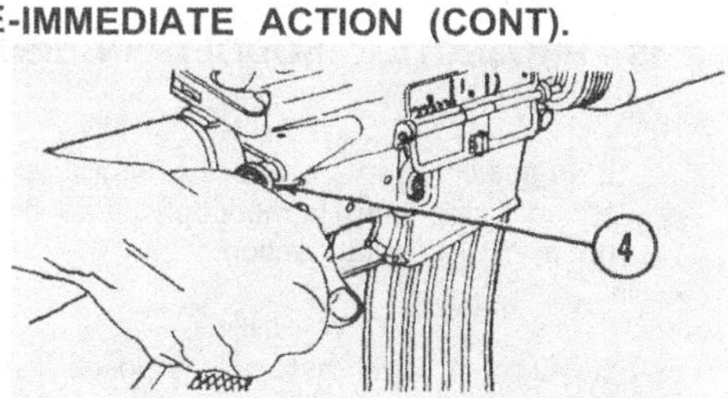

4 Tap forward assist assembly (4) (MI 6A1 only).

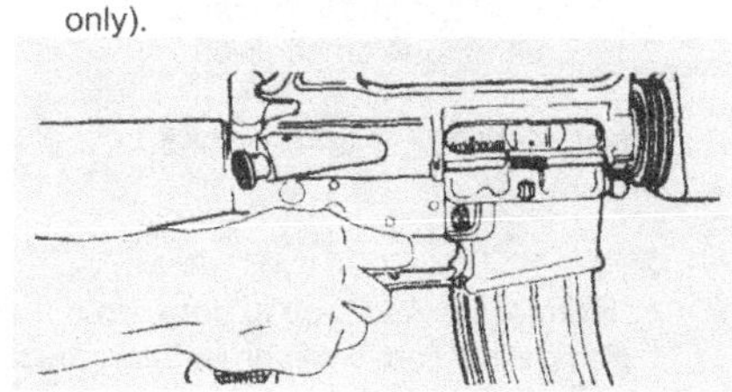

5 Now shoot. If it won't fire, look for the trouble and apply remedial action (p 2-31).

2-17. OPERATING PROCEDURE - REMEDIAL ACTION.

WARNING

If rifle stops firing with a live round in the chamber of a hot barrel, remove the round fast. However, during training, if you cannot remove it within 10 seconds, wait 15 minutes with the rifle pointing in a safe direction This way you won't get hurt by a possible ammunition cook-off, which could happen 10 seconds after contact with a hot chamber. Clear the rifle.

1 Check for jammed cartridge case.

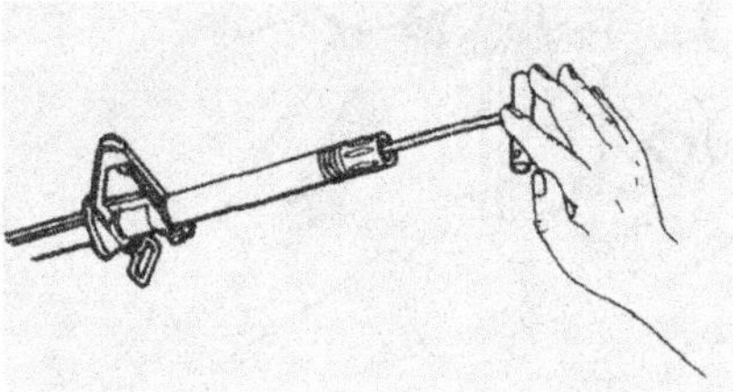

2 If a cartridge case is in the chamber, tap out with a cleaning rod.

NOTE

If your rifle still fails to fire, check troubleshooting.

2-18. PROJECTILE LODGED IN BORE.

WARNING

If an audible "pop" or reduced recoil is experienced during firing, immediately cease fire. Do not apply immediate action.

WARNING

Be sure bolt carrier assembly is closed (forward). If barrel is hot, wait 15 minutes for barrel and extension assembly to cool so you won't be hurt by an ammunition cook-off.

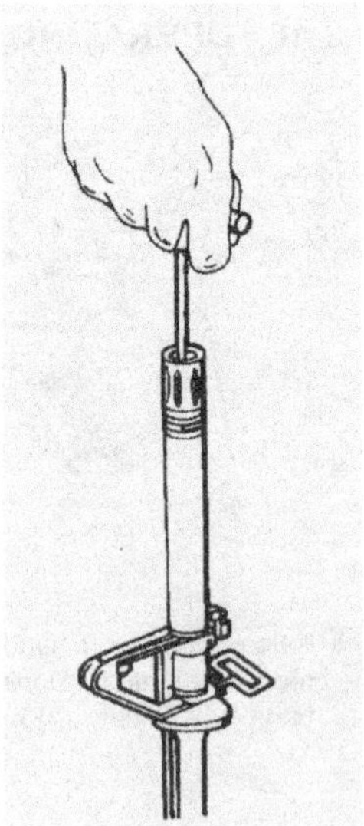

1 Remove cartridge magazine, lock bolt, and return charging handle (p 2-33).

CAUTION

If projectile is lodged in bore, do not attempt to remove it. Turn weapon in to organizational maintenance.

2 Visually inspect and/or inserf a cleaning rod into bore to ensure there is not a projectile lodged in bore.

2-19. OPERATING PROCEDURES - CLEARING YOUR RIFLE.

WARNING

To avoid accidental firing, always look into chamber after clearing weapon to make sure it does not contain a round.

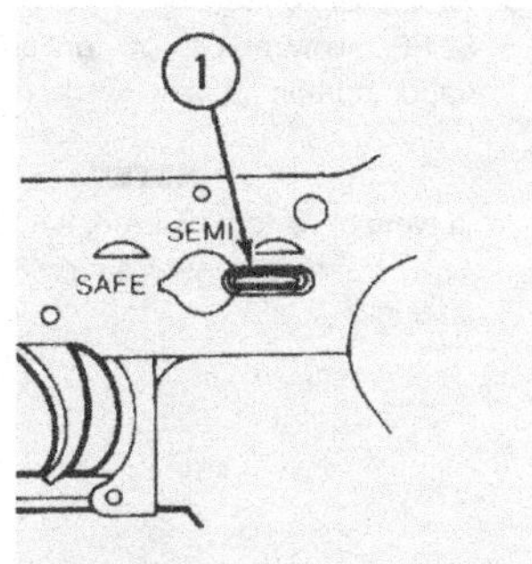

1 Place selector lever (1) on SAFE.

NOTE

If weapon is not cocked, lever cannot be pointed toward SAFE.

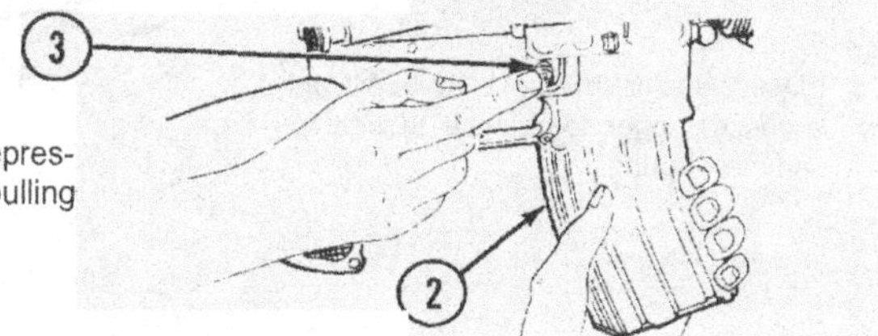

2 Remove cartridge magazine (2) by depressing magazine catch button (3) and pulling cartridge magazine (2) down.

2-19. OPERATING PROCEDURES - CLEARING YOUR RIFLE (CONT).

3 To lock bolt open, pull charging handle assembly (4) rearward, press bottom of bolt catch (5), and allow bolt to move forward until it engages bolt catch. Return charging handle assembly (4) forward.

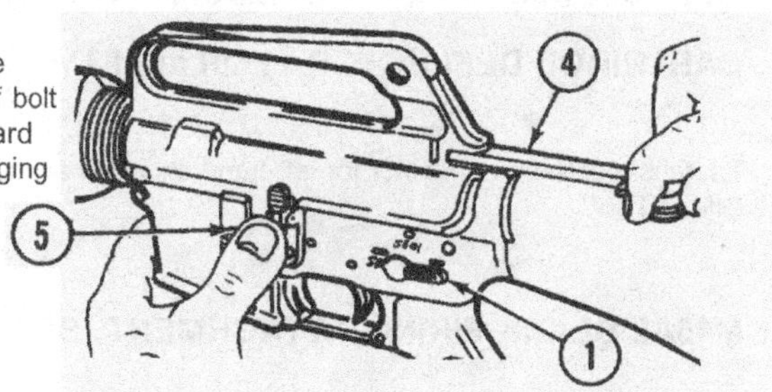

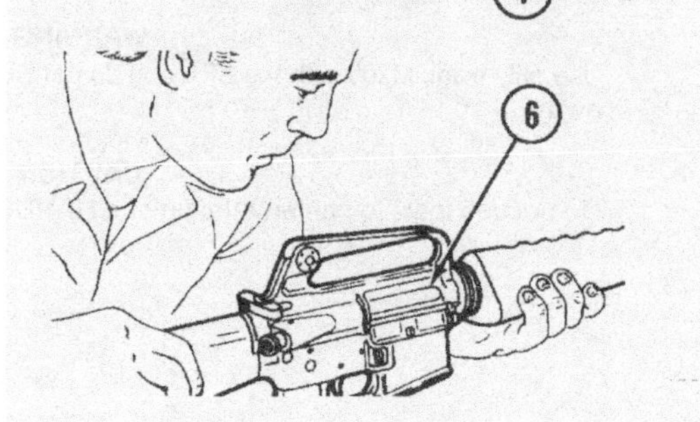

NOTE

Ensure that selector lever (1) is on SAFE.

4 Check receiver and chamber (6) to ensure these areas contain no ammunition.

5 With selector lever (1) pointing toward SAFE, allow bolt to go forward by pressing upper portion of bolt catch (5).

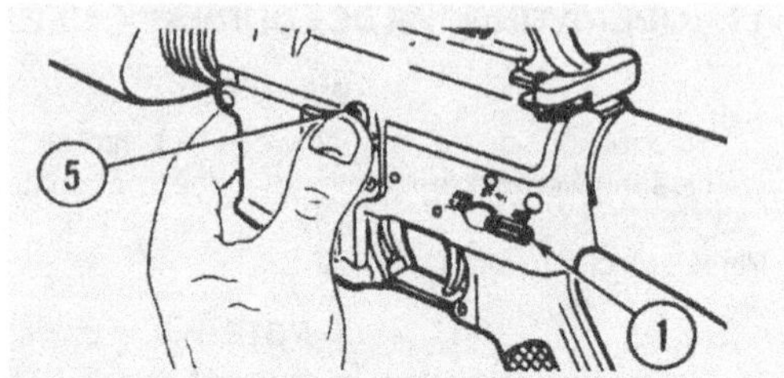

NOTE

If weapon is to be stored, it should be dry fired to release tension on hammer spring.

6 Place selector lever (1) on SEMI and squeeze trigger to release tension on hammer spring.

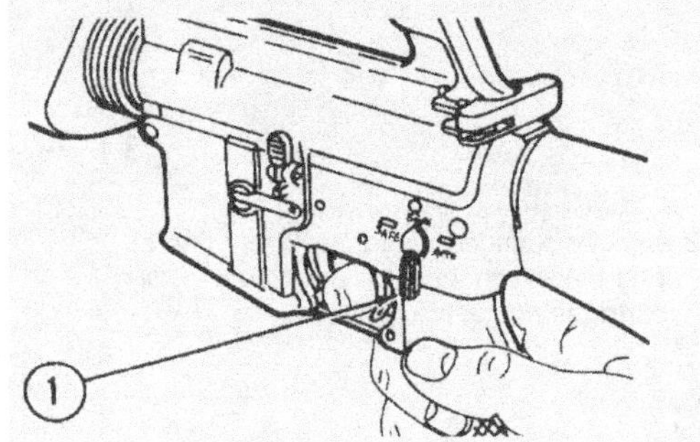

2-35

2-20. OPERATION OF AUXILIARY EQUIPMENT.

CARTRIDGE DEFLECTOR - FOR “LEFTYS”

There is a cartridge deflector for left-hand shooters available from your Training Aids Service Office (TASO).

M15A2 BLANK FIRING ATTACHMENT (BFA)

WARNING

Use only blank M200 with the BFA and do not fire directly at anyone less than 20 feet away

CAUTION

Do not use tools to tighten attachment, HANDS ONLY

NOTE

After 50 rounds, check to see If BFA is still tight. Make sure to clean carbon buildup after each training exercise.

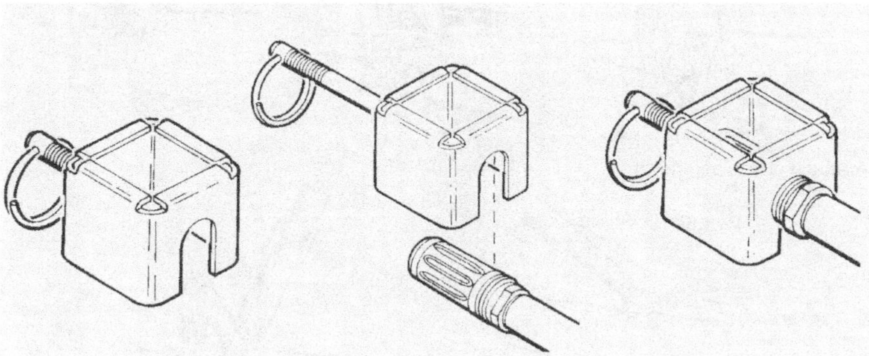

1 Unscrew and slide all the way back.

2 Hook behind first groove of flash suppressor.

3 Slide into flash suppressor and hand tighten.

2-20. OPERATION OF AUXILIARY EQUIPMENT (CONT).

TOP SLING ADAPTER

1 **Remove rifle sling from rifle.**

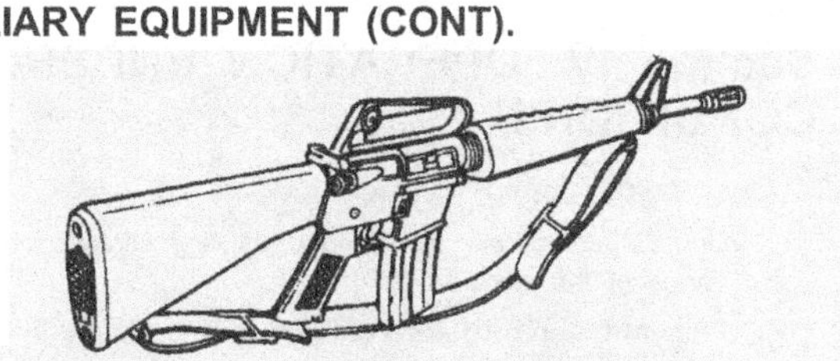

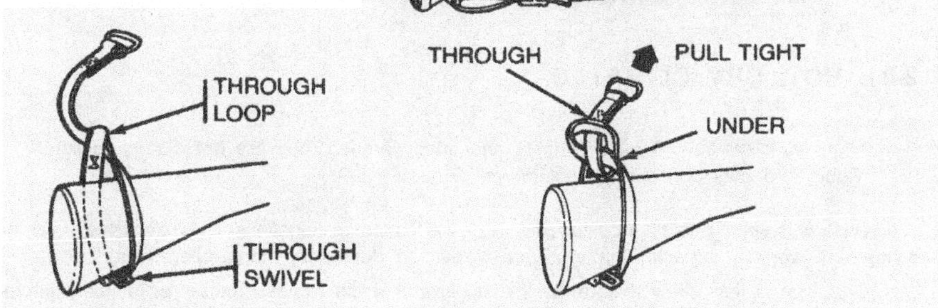

2 Work adapter sling through swivel and through loop.

3 Work buckle under and then through loop,

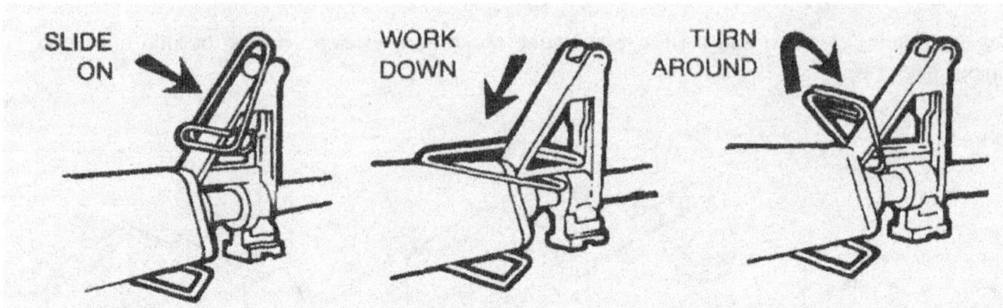

4 Attach clamp as shown.

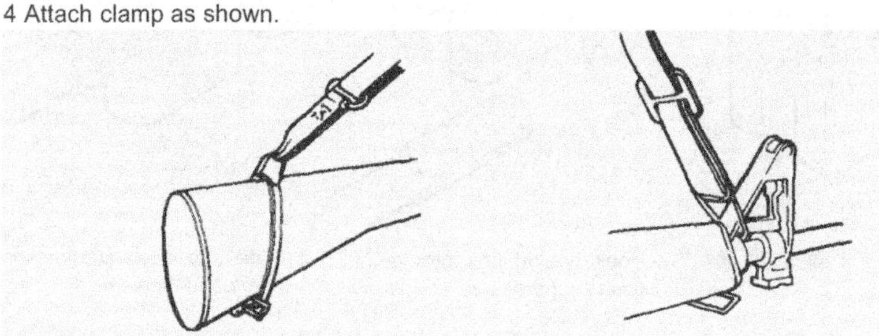
5 Attach rifle sling to top sling adapter as shown.

Section IV. OPERATION UNDER UNUSUAL CONDITIONS

NOTE

Unusual conditions are defined as any climatic condition requiring special maintenance of the weapon.

Perform the maintenance outlined for the climate that most applies to your operational area.

2-21. HOT, DRY CLIMATES.

NOTE

Hot, dry climates are usually dusty and sandy areas, They are hot during daylight hours and cool during the night hours,

a. Dust and sand will get into the rifle and will cause malfunctions and excessive wear on component working surfaces through abrasive action during the firing operations,

b. Corrosion is less likely to form on metal parts in a dry climate; therefore, lubricate internal working surfaces only with a small amount of cleaner, lubricant and preservative (CLP) (item 1, app D) (always shake CLP prior to use). Do not lubricate external parts of the rifle, Doing so will only collect dust and sand, making the rifle difficult to keep clean. Do not lubricate internal components of the magazines,

c. Using Additional Authorization List (AAL) equipment, i.e., protective cap and spare magazine protective bags and overall rifle protective cover (item 3, app D) will help keep dust and sand from getting into the rifle. Use these items as the tactical situation allows. As a minimum effort to keep dust and sand out of the rifle, keep the ejection port cover closed, a cartridge magazine installed in the rifle, and a muzzle cap on the muzzle.

NOTE

More firing, remove the protective cap and keep for later use. However, it is not dangerous to fire the rifle with the protective cap. The cap will blow off when the first round is fired and may be lost.

2-22. HEAVY RAIN AND FORDING. Perform maintenance according to the appropriate climatic condition. Use AAL equipment and expendable items to protect the rifle, Always keep the rifle dry. Using the protective cap will help keep water out of the barrel. Always drain any water from the barrel before firing. Dry the bore with a swab (item 5, app D) and cleaning rod, if necessary

2-23. EXTREMELY COLD CLIMATES.

a. To operate the rifle in extremely cold climates, depress the trigger guard plunger and open the trigger guard. This makes it easier to operate the trigger when you are wearing arctic mittens.

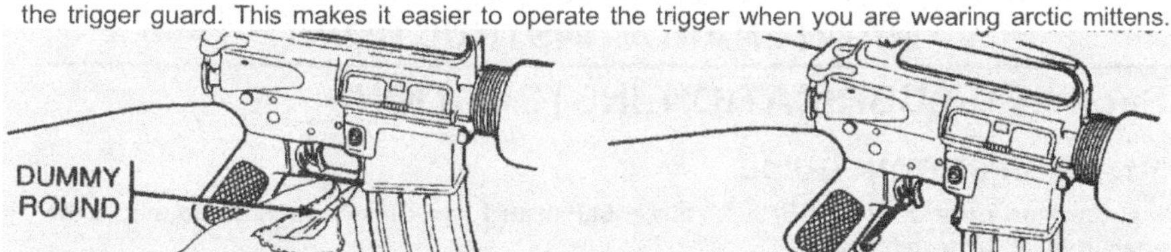

b. Clean and lubricate the rifle inside a warm room. The rifle should be at room temperature, if possible.

(1) Apply a light coat of CLP (item 1, app D) to all functional parts,

NOTE

Always shake CLP to obtain a good mixture before applying.

(2) To prevent freezing, keep the weapon covered when moving from a warm to a cold area to allow gradual cooling.

(3) Always keep the weapon dry.

(4) Hand function the weapon approximately every 30 minutes to help prevent freezing of components.

(5) Do not lay a hot weapon in, snow or ice.

(6) Keep the ammunition dry. Moisture will cause malfunctions. Do not lubricate the ammunition.

(7) Using AAL equipment, i.e., protective cap and protective bag and protective cover (item 3, app D) will help protect your rifle. Use them whenever the tactical condition is suitable. Always keep snow out of the barrel bore. Clean barrel bore with swab (item 5, app D) and cleaning rod, if necessary, before firing.

2-24. HOT, WET CLIMATES.

a. Perform normal maintenance as outlined in the PMCS table (see page 2-4).

b. Perform maintenance more frequently. Inspect hidden surfaces of the bolt carrier assembly, upper receiver and barrel assembly, and lower receiver and extension assembly for corrosion. If corrosion is found, clean and lubricate.

c. To help prevent corrosion, remove handprints with a wiping rag (item 4, app D). Dry and lubricate lightly with CLP (item 1, app D).

d. Check ammunition and cartridge magazines frequently for corrosion. Clean using CLP (item 1, app D) and wipe dry with wiping rag (item 4, app D),

e. Use appropriate AAL equipment and expendable items for protection when the tactical conditions allow.

Section V. NUCLEAR, BIOLOGICAL, AND CHEMICAL (NBC).

General procedures can be found in FM 3-87, FM 21-40, and TM 3-220. Refer to page 3-13.

CHAPTER 3
MAINTENANCE INSTRUCTIONS

Section I. LUBRICATION INSTRUCTIONS

3-1. LUBRICATION GUIDE.

a. *Cleaner, Lubricant and Preservative.* CLP (item 1, app D) is the lubricant to use on the weapon at all temperatures.

b. Lightly *Lubricate.* A film of oil barely visible to the eye.

c. *Generously Lubricate.* Heavy enough so that it can be spread with the finger.

NOTE

These lubrication instructions are mandatory.

Section II. TROUBLESHOOTING PROCEDURES

3-2. INTRODUCTION.

a. The table lists the common malfunctions which you may find during the operation or maintenance of the rifle. You should perform the teats/inspections and corrective actions in the order listed.

b. This manual cannot list all malfunctions that may occur, nor all tests or inspections and corrective actions. If a malfunction is not listed or is not corrected by listed corrective actions, notify organizational maintenance.

TROUBLESHOOTING

MALFUNCTION
TEST OR INSPECTION
CORRECTIVE ACTION

1. WEAPON WILL NOT FIRE.

Step 1. Check to see if selector lever (1) is in SAFE position.
Place in SEMI or AUTO position.

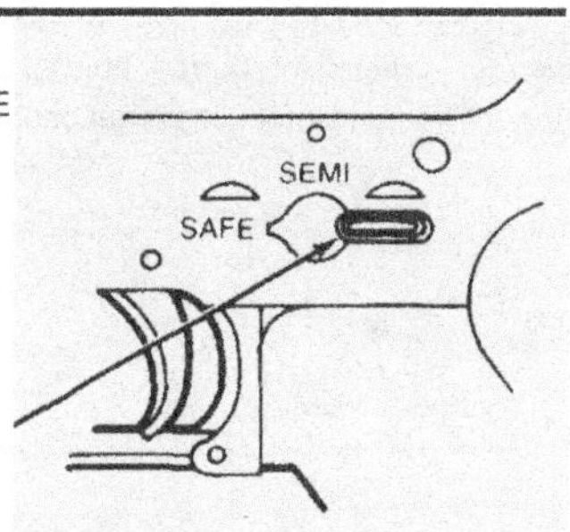

Step 2. Check for improper assembly of firing pin.
Assemble correctly (p 3-43).

TROUBLESHOOTING (CONT)

MALFUNCTION
TEST OR INSPECTION
CORRECTIVE ACTION

1. **WEAPON WILL NOT FIRE (CONT).**

Step 3. Check for too much oil in firing pin recess (2).
Wipe off.

Step 4. Check for defective ammunition.
Remove and discard (p 4-1).

Step 5. Check for too much carbon on firing pin.
Clean (p 3-34).

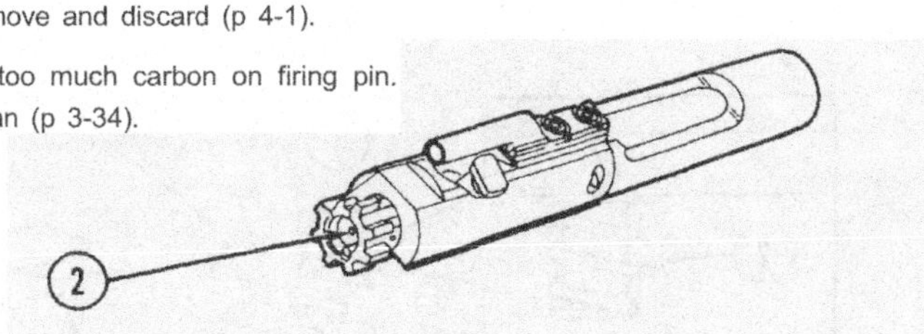

TROUBLESHOOTING (CONT)

MALFUNCTION
TEST OR INSPECTION
CORRECTIVE ACTION

2. BOLT WILL NOT UNLOCK.

Step 1. Check for dirty bolt (1).
Clean (p 3-34).

Step 2. Check for burred bolt(1).
Notify organizational maintenance.

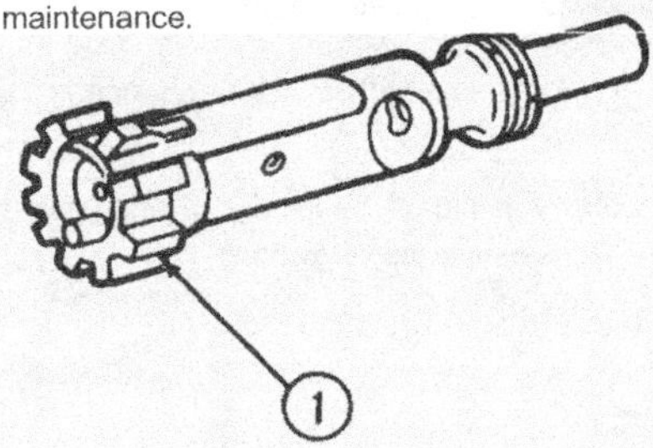

TROUBLESHOOTING (CONT)

MALFUNCTION
TEST OR INSPECTION
CORRECTIVE ACTION

3. WEAPON WILL NOT EXTRACT.

Step 1. Check for stuck cartridge or cartridge case in chamber.
Remove stuck cartridge or cartridge case (p 2-31)

Step 2. Check for broken cartridge extractor (1) or extractor spring assembly (2)
Replace bolt assembly and notify organizational maintenance.

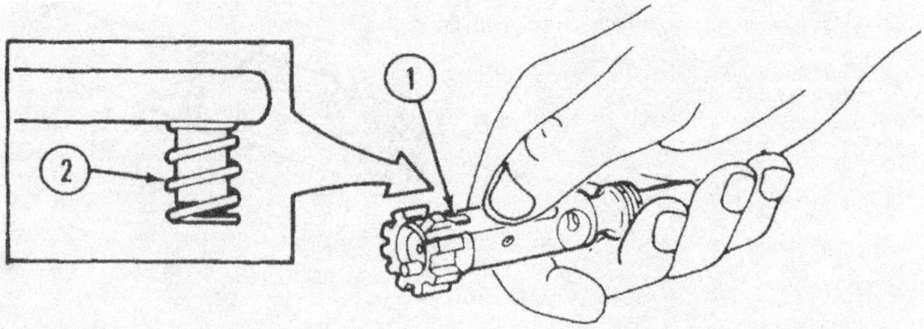

TROUBLESHOOTING (CONT)

MALFUNCTION
TEST OR INSPECTION
CORRECTIVE ACTION

3. WEAPON WILL NOT EXTRACT (CONT).

Step 3. Check for dirty or corroded ammunition.

a. Clean dirty ammunition with dry cloth
b. Corroded ammunition must be returned to unit supply or ammunition personnel.

Step 4. Check for carbon in chamber.

Clean (p 3-31).

Step 5. Check for carbon in cartridge extractor opening (3) on bolt assembly or cartridge extractor lip (4).

Clean (p 3-35).

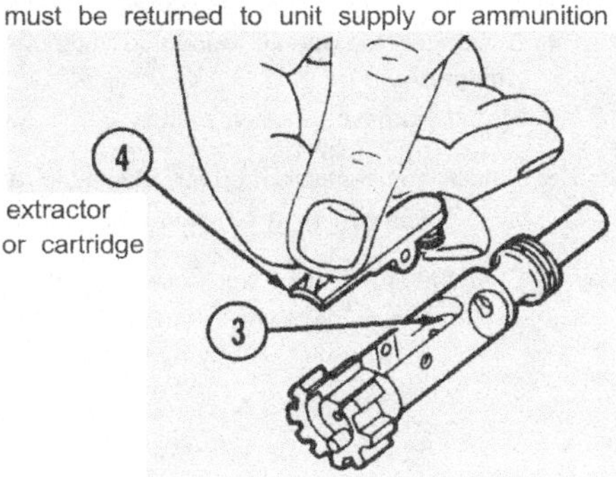

TROUBLESHOOTING (CONT)

MALFUNCTION
TEST OR INSPECTION
CORRECTIVE ACTION

4. WEAPON WILL NOT EJECT CARTRIDGE CASE.

Step 1. Check for broken ejector spring.

Report to unit armorer for repair.

Step 2. Bolt assembly is installed upside down.

The bolt cam pin is worn or the bolt cam pin hole is defective and allows for improper assembly. Disassemble and assemble correctly. Report the deficiency to unit armorer.

5. WEAPON WILL NOT FEED.

Step 1. Check for dirty or corroded ammunition.

a. Clean dirty ammunition with dry cloth.
b. Corroded ammunition must be returned to unit supply or ammunition personnel.

Step 2. Check for dirty cartridge magazine.

Clean (p 3-26).

TROUBLESHOOTING (CONT)

MALFUNCTION
TEST OR INSPECTION
CORRECTIVE ACTION

5. WEAPON WILL NOT FEED (CONT).

Step 3. Check for defective cartridge magazine(l).
Replace (app C).

Step 4. Check for too many rounds in cartridge magazine (1).
Remove excess rounds.

Step 5. Check for restricted buffer assembly action.
Remove (p 3-23) and clean (p 3-39).

Step 6. Cartridge magazine not fully seated.
Adjust magazine catch (p 2-10).

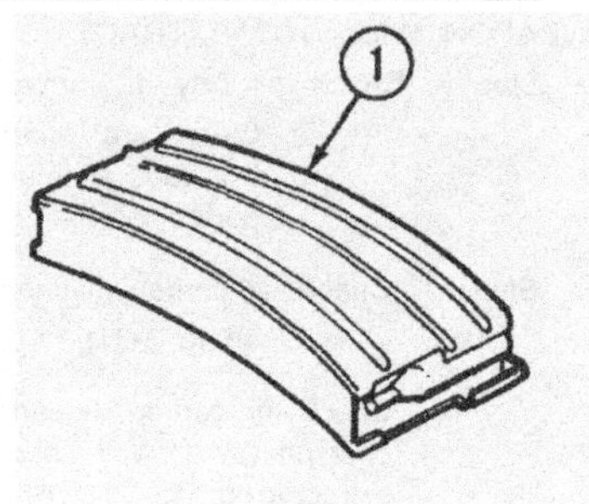

TROUBLESHOOTING (CONT)

MALFUNCTION
TEST OR INSPECTION
CORRECTIVE ACTION

6. ROUND WILL NOT CHAMBER.

Step 1. Check for dirty or corroded ammunition.
a. Clean dirty ammunition with dry cloth.
b. Corroded ammunition must be returned to unit supply or ammunition personnel.

Step 2. Check for damaged ammunition.
Replace (p 4-1).

Step 3. Check for carbon in chamber (1).
Clean (p 2-14).

TROUBLESHOOTING (CONT)

MALFUNCTION
TEST OR INSPECTION
CORRECTIVE ACTION

7. BOLT WILL NOT LOCK.

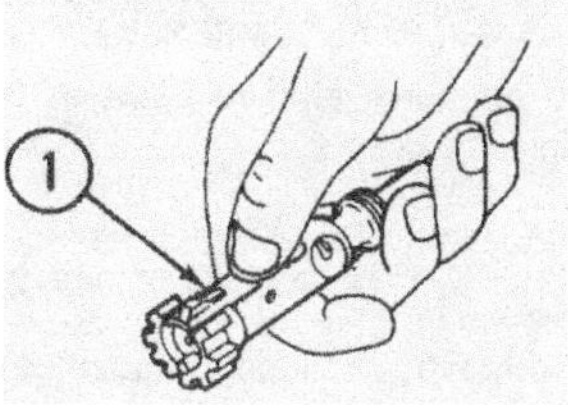

Step 1. Check for dirt, corrosion, or carbon buildup in barrel locking lugs
Clean (p 3-35).

Step 2. Check for frozen cartridge extractor (1).
Remove (p 3-21) and clean (p 3-34).

step 3. Check for restricted buffer assembly action.
Remove (p 3-23) and clean (p 3-39).

Step 4. Restricted movement of bolt carrier assembly.
Remove (p 3-17), clean (p 3-34), and lubricate.

NOTE

When putting bolt back in, make sure gas tube fits into bolt carrier assembly and moves freely.

TROUBLESHOOTING (CONT)

MALFUNCTION
TEST OR INSPECTION
CORRECTIVE ACTION

8. WEAPON HAS SHORT RECOIL.

Step 1. Gaps in bolt rings (not staggered).
Stagger ring gaps (p 3-42).

Step 2. Carbon or dirt in key and carrier assembly or on outside of gas tube.
Clean (p 3-34 and 3-30).

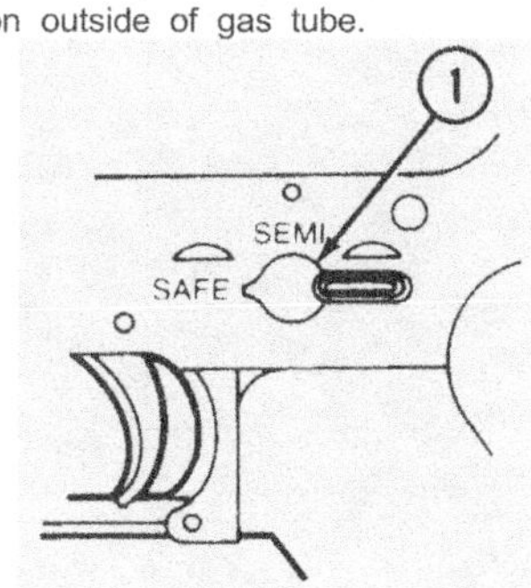

9. BOLT FAILS TO LOCK AFTER LAST ROUND.

Step 1. Check for dirty or corroded bolt catch.
Clean (p 3-38).

Step 2. Check for faulty cartridge magazine.
Replace (app B).

10, SELECTOR LEVER BINDS.

Check for not enough lubrication on selector lever (l).
Lubricate with CLP (item 1, app D).

TROUBLESHOOTING (CONT)

MALFUNCTION
TEST OR INSPECTION
CORRECTIVE ACTION

11. BOLT CARRIER ASSEMBLY "HUNG UP."

Check for round jammed between bolt and charging handle.

WARNING

Keep clear of muzzle.

a. Remove cartridge magazine.

CAUTION

After round is removed, bolt is under tension.

b. Hold charging handle assembly (1) back and bang rifle buttstock (2) on ground.

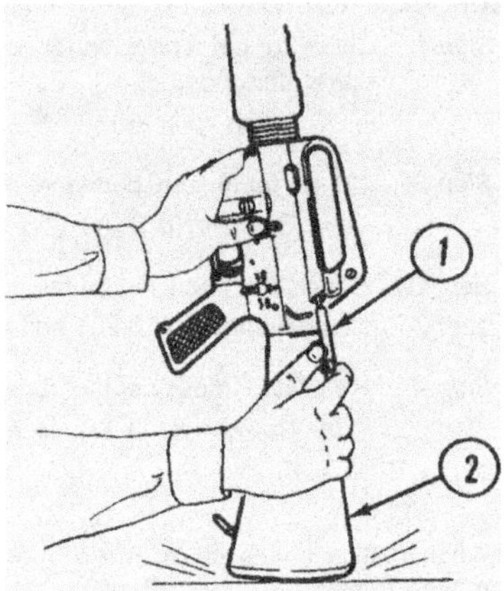

TROUBLESHOOTING (CONT)

MALFUNCTION
TEST OR INSPECTION
CORRECTIVE ACTION

11. BOLT CARRIER ASSEMBLY "HUNG UP" (CONT)

c. While bolt is held to rear, push charging handle assembly (1) forward. Round should fall through magazine well (3).

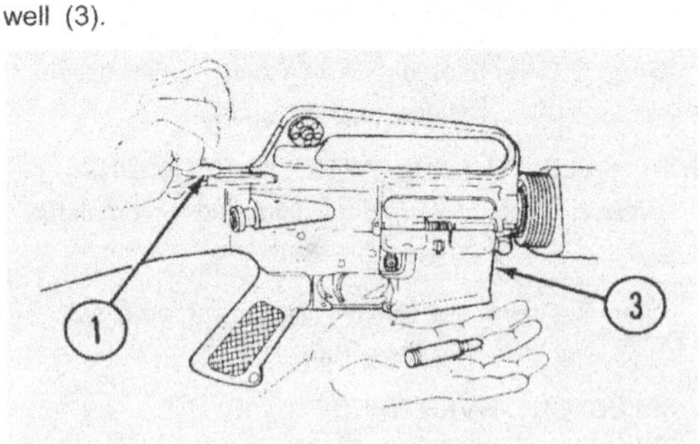

TROUBLESHOOTING (CONT)

MALFUNCTION

TEST OR INSPECTION

CORRECTIVE ACTION

12. CAN'T SEE TRITIUM SIGHT POST.

WARNING

RADIATION HAZARD

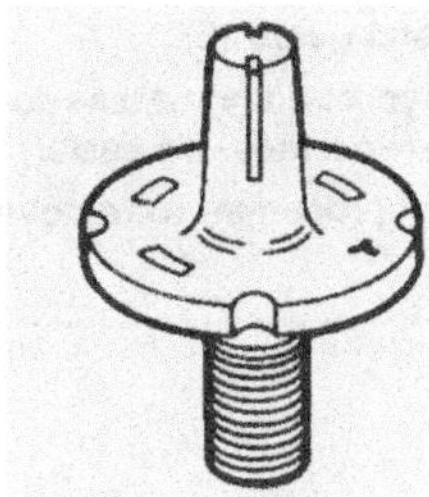

In the event there is no illumination, notify the local Radiological Protection Officer. Do not attempt to repair or replace the sight in the field! if skin contact is made with any area contaminated with tritium, immediately wash with nonabrasive soap and water.

Step 1. Luminous element broken or missing.

Turn in to unit armorer. Sights must be turned in as radioactive waste,

Step 2. Check for alinenaent of luminous elements.

Rotate front sight post up or down one click

Step 3. Luminous element dirty.

Clean using swab and CLP (item 1, app D).

Section III. MAINTENANCE PROCEDURES

3-3. FIELD-STRIPPING M16/M16A1 RIFLE.

CLEARING RIFLE

WARNING

To avoid accidental firing, be sure rifle is , clear (p 2-33).

Pull back charging handle (1) and check chamber (2). Place selector lever on SAFE.

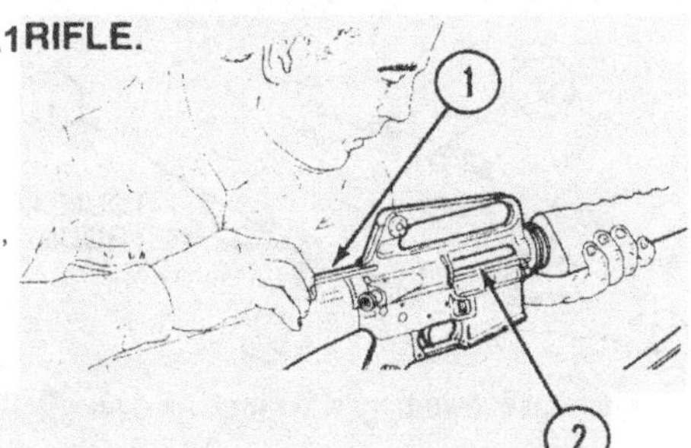

SLING

Remove sling (1).

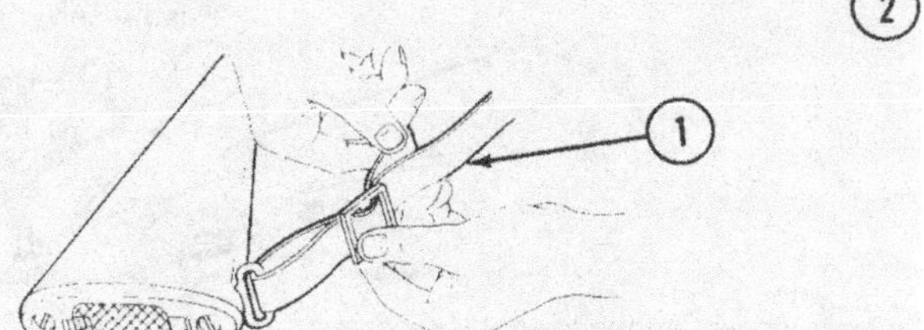

HAND GUARDS - THE BUDDY SYSTEM

1 Remove and clean hand guards only if dirt and corrosion can be seen through hand guard vent holes.

2 Place the weapon on the buttstock (1) with one hand gripping the stock and the other gripping the lower end of the hand guard (2).

3 While your buddy presses down on the slip ring (3) (tell him to use two hands - it's easier), pull the hand guard (2) free.

4 Repeat procedure to remove other hand guard.

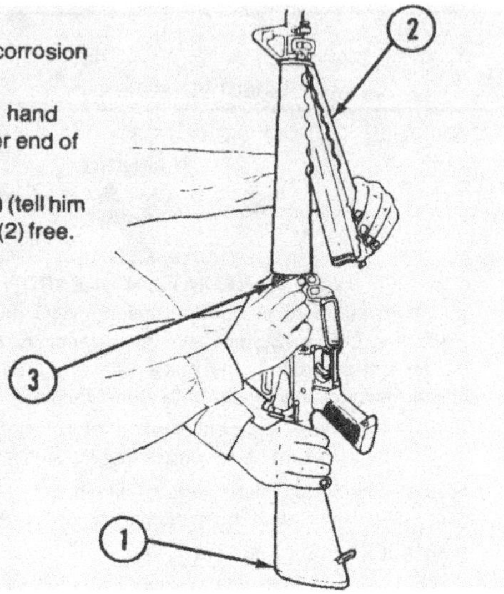

3-3. FIELD-STRIPPING M16/M16A1 RIFLE (CONT).

UPPER AND LOWER RECEIVERS

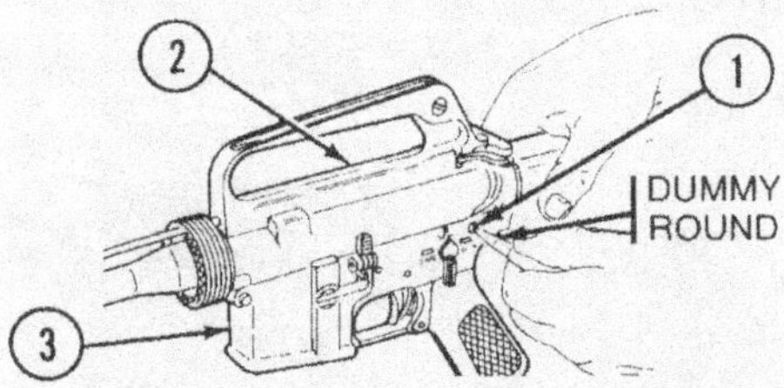

1 Push takedown pin (1) as far as it goes.

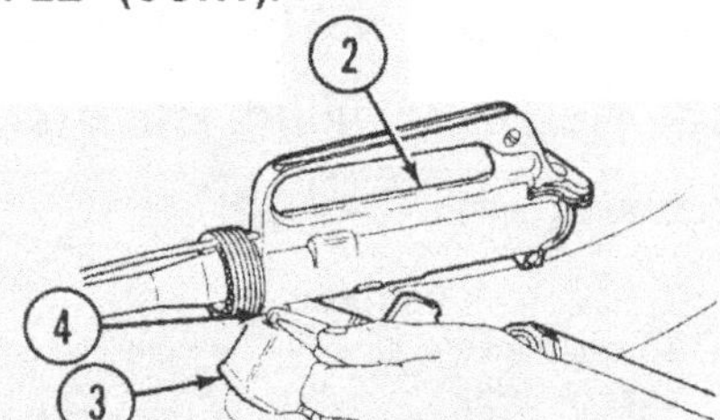

2 Pivot upper receiver (2) from lower receiver (3).

3 Push receiver pivot pin (4).

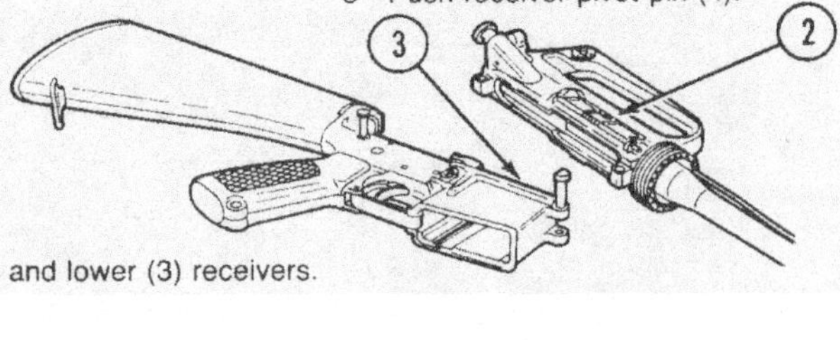

4 Separate upper (2) and lower (3) receivers.

CHARGING HANDLE ASSEMBLY AND BOLT CARRIER ASSEMBLY

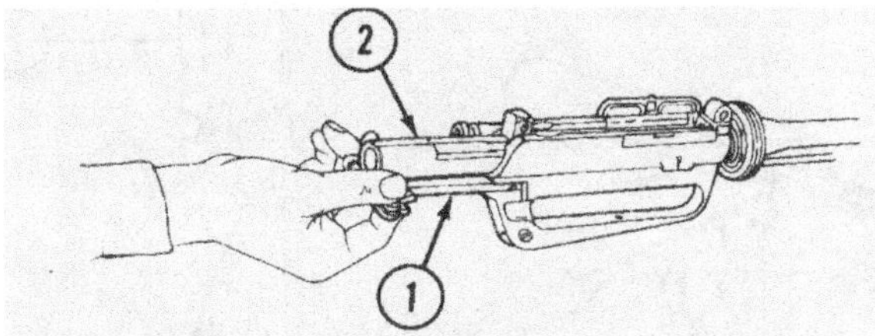

1 Pull back charging handle assembly (1) and bolt carrier assembly (2).

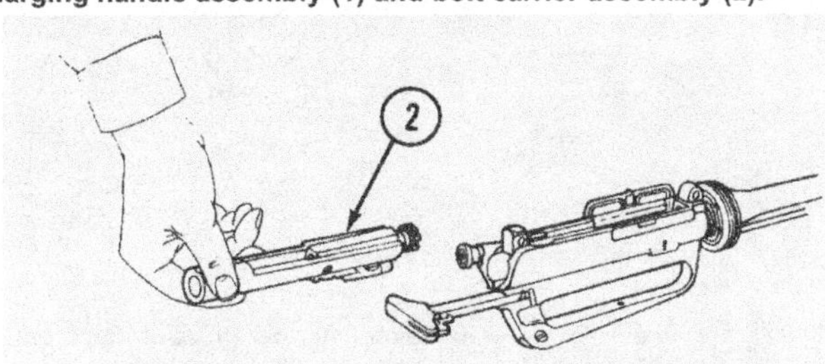

2 Remove bolt carrier assembly(2)

3-3. FIELD-STRIPPING M16/M16A1 RIFLE (CONT).

CHARGING HANDLE ASSEMBLY AND BOLT CARRIER ASSEMBLY (CONT)

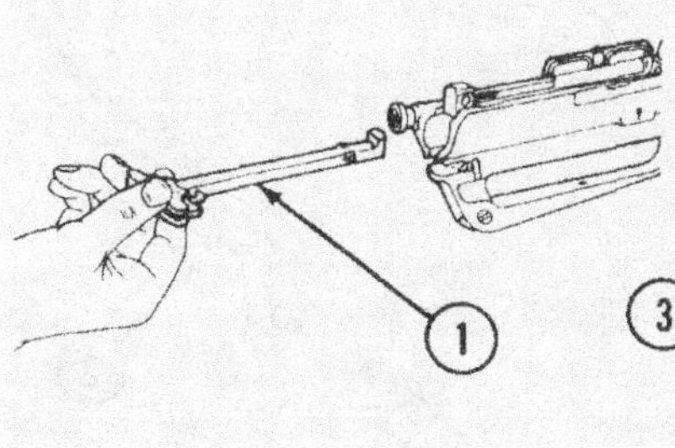

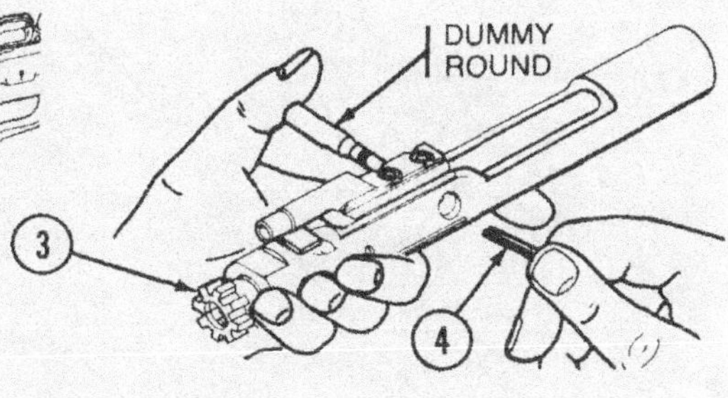

3 Pull charging handle assembly (1) back and down and remove it.

4 Move bolt assembly (3) forward to unlocked position and remove firing pin retaining pin (4). Do not open or close split end of firing pin retaining pin (4).

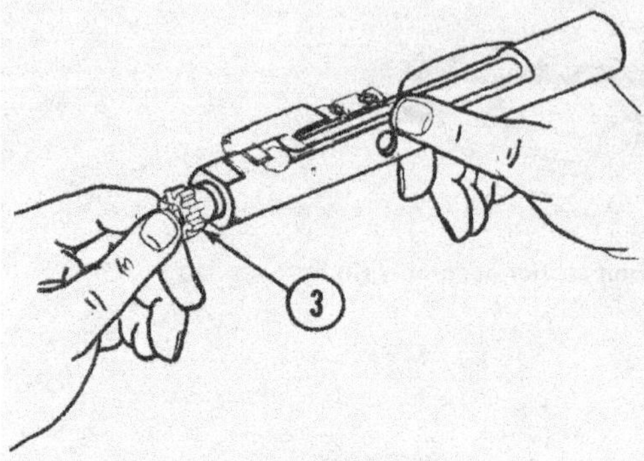

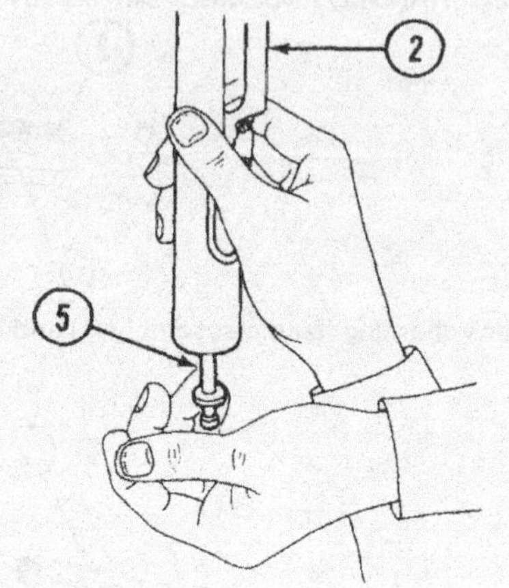

5 Push in on bolt assembly (3) to put in locked position.

6 Catch firing pin (5) as it drops out of rear of bolt carrier assembly (2).

3-3. FIELD-STRIPPING M16/M16A1 RIFLE (CONT).

CHARGING HANDLE ASSEMBLY AND BOLT CARRIER ASSEMBLY (CONT)

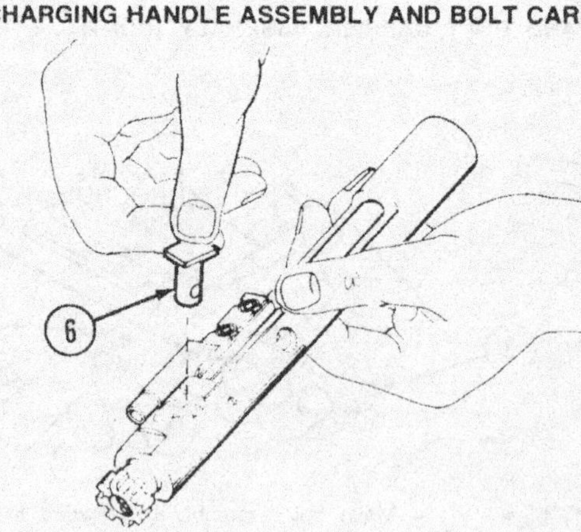

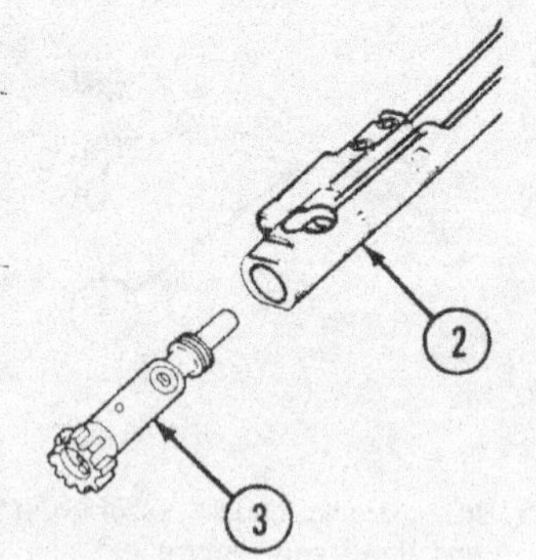

7 Give bolt cam pin(6) a 1/4 turn and lift out.

8 Remove bolt assembly (3) from bolt carrier assembly (2).

NOTE

Disassemble extractor and spring assembly only when dirty or damaged.

9 Press top of extractor (7) to check that spring works.

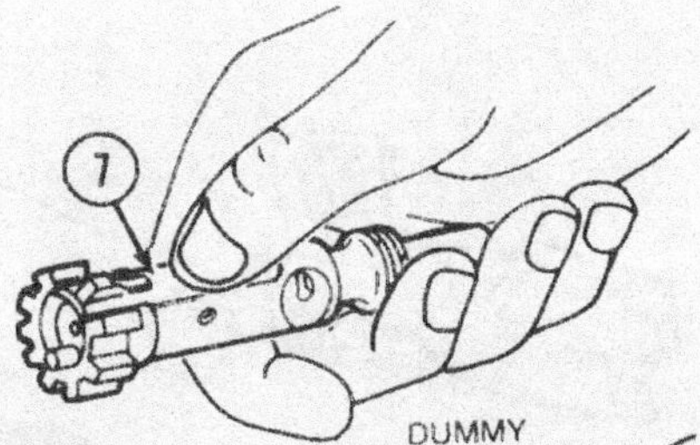

10 **Use round (5) to remove extractor pin (8).**

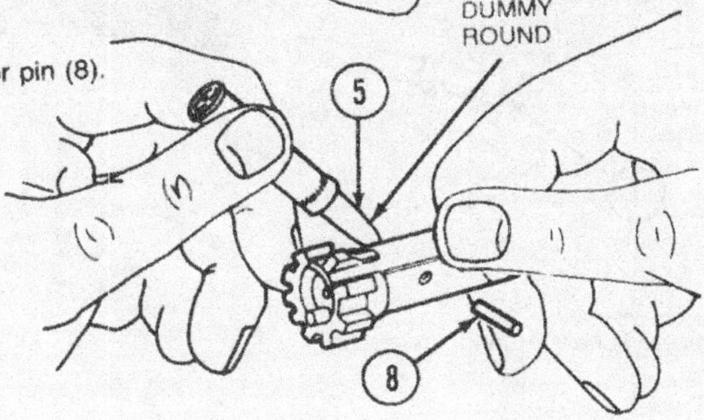

3-3. FIELD-STRIPPING M16/M16A1 RIFLE (CONT).

CHARGING HANDLE ASSEMBLY AND BOLT CARRIER ASSEMBLY (CONT)

CAUTION

Do not separate insert from spring assembly (9).

11 Remove extractor (7) and spring assembly (9). Do not remove spring assembly (9) from extractor (7).

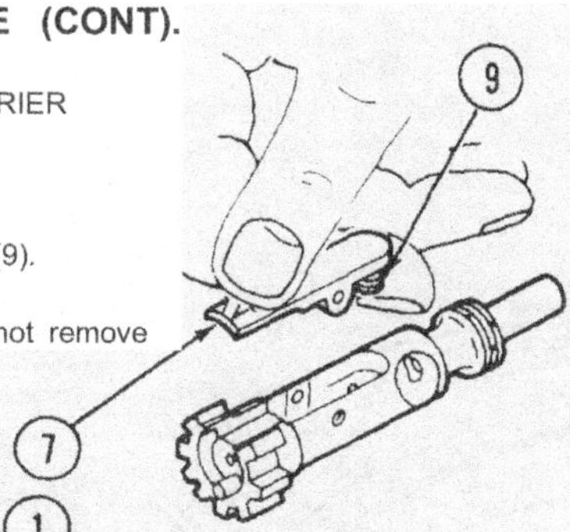

LOWER RECEIVER AND EXTENSION ASSEMBLY

1 Press in buffer (1) and depress retainer (2) to release buffer.

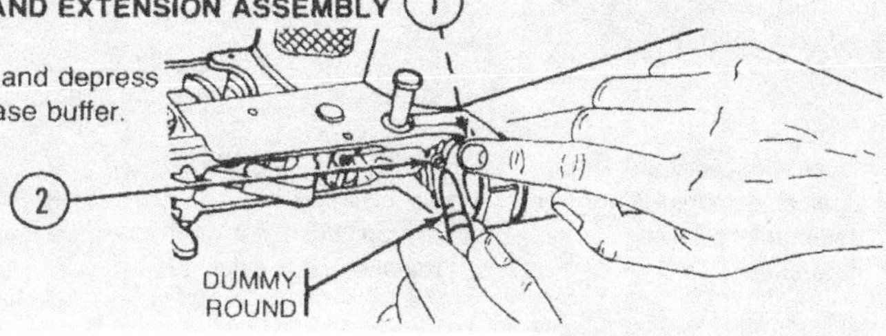

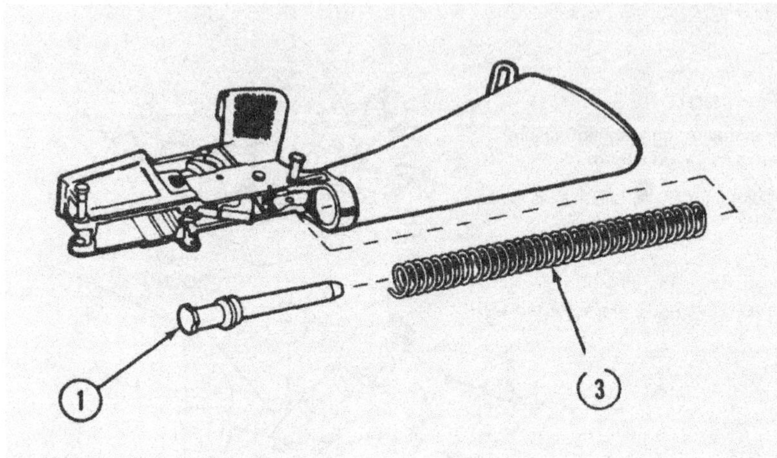

2 Remove buffer (1) and spring (3).

CAUTION
No further disassembly is allowed.

3-4. FIELD-STRIPPING CARTRIDGE MAGAZINE.

NOTE
Disassemble only if cartridge magazine is dirty

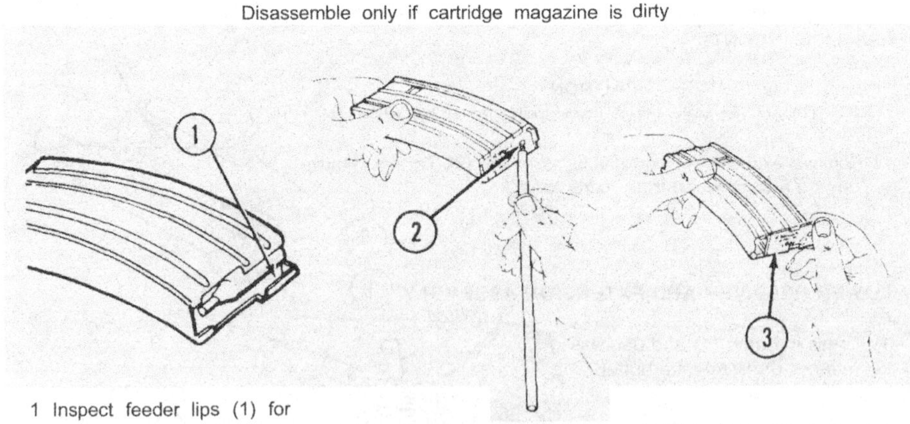

1 Inspect feeder lips (1) for wear or damage. If worn or damaged, replace magazine.

2 Use cleaning rod section to pry up, push out, and release base catch (2).

3 Slide base (3) from cartridge magazine,

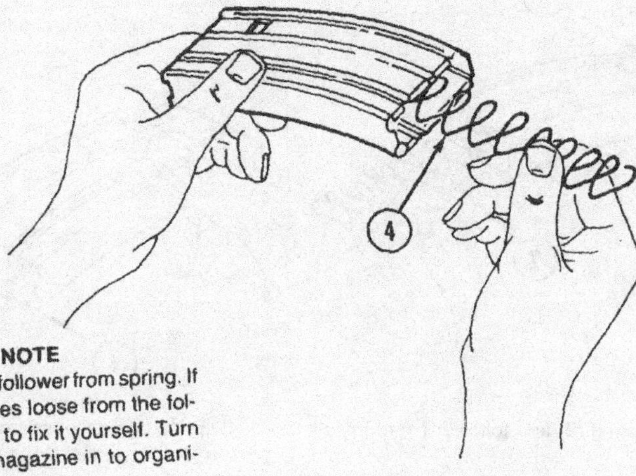

NOTE

Do not remove follower from spring. If the spring comes loose from the follower, don't try to fix it yourself. Turn the cartridge magazine in to organizational maintenance.

4 Jiggle spring(4) and follower to remove.

3-5. MAINTENANCE OF CARTRIDGE MAGAZINE.

CLEANING, INSPECTION, AND REASSEMBLY

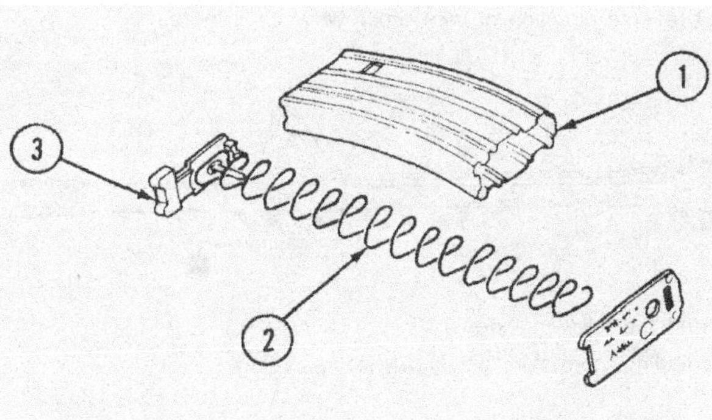

1 Clean and lubricate, Use swab (item 5, app D) to wipe dirt from tube (1), spring (2), and follower (3). Lightly lubricate spring (2) and follower (3) with CLP (item 1, app D).

2 Inspect spring (2) and follower (3) for damage. If parts are damaged, replace cartridge magazine.

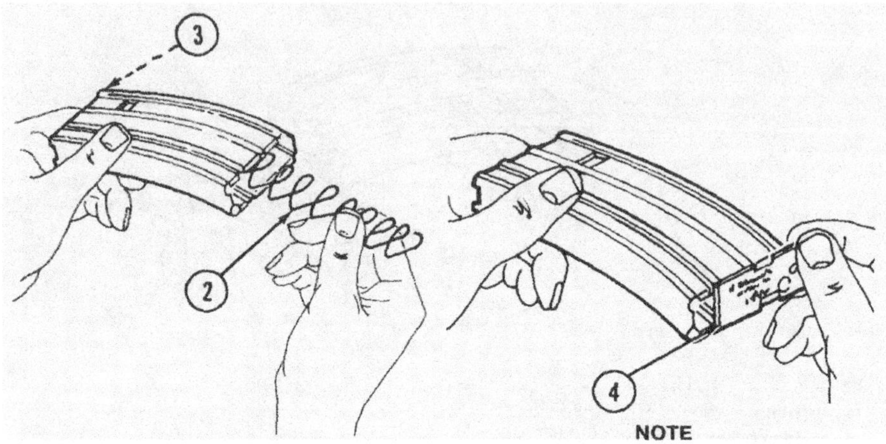

3 Jiggle spring (2) and follower (3) to install.

NOTE

If spring comes loose from follower, turn in the pieces. Don't try to fix it yourself.

4 Make sure printing is on the outside of the base. Slide base (4) under all four tabs.

3-6. MAINTENANCE OF UPPER RECEIVER AND BARREL ASSEMBLY.

CLEANING

NOTE

Don't reverse direction of bore brush while it's in bore.

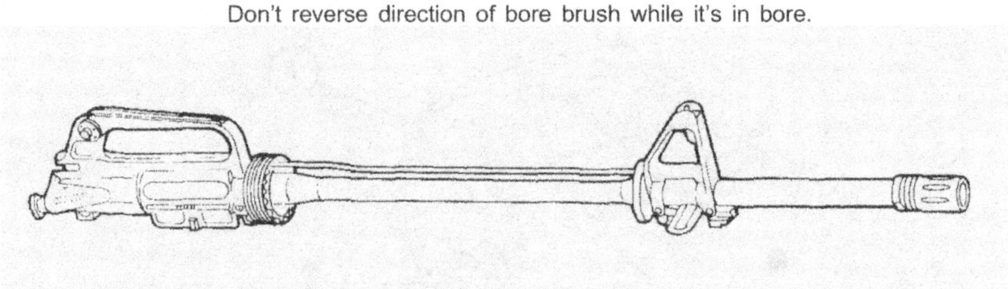

Use CLP (item 1,app D) on the following areas:

a. All areas of powder fouling, corrosion, dirt and rust

b. Bore and chamber.

c. Upper receiver and barrel assembly locking lugs.

d. Gas tube.

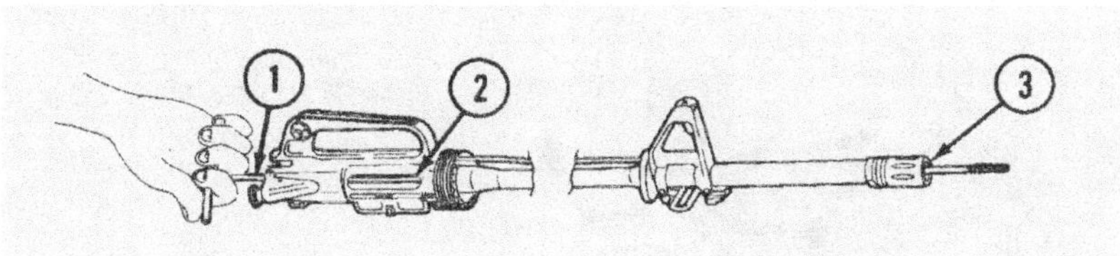

1 Use cleaning rod, bore brush, and CLP (item 1, app D). Run rod (1) through chamber (2) and flash suppressor (3).

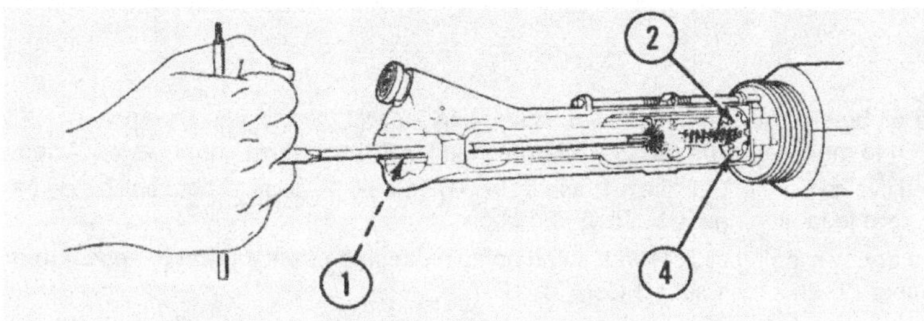

2 Install chamber brush on cleaning rod (1), dip in CLP (item 1, app D), and insert in chamber (2) and locking lugs (4). Clean by pushing and twisting cleaning rod.

3-6. MAINTENANCE OF UPPER RECEIVER AND BARREL ASSEMBLY (CONT).

CLEANING (CONT)

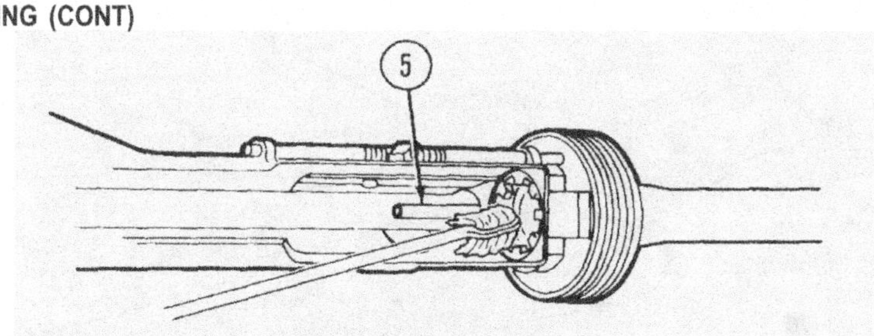

NOTE

Gas tubes will discolor from heat. Do not attempt to remove discoloration.

Use a wornout bore brush to perform the following step. This procedure ruins the bore brush.

3 Use a bore brush to clean outside surface of protruding gas tube (5) (get sides and bottom from bottom of upper receiver).

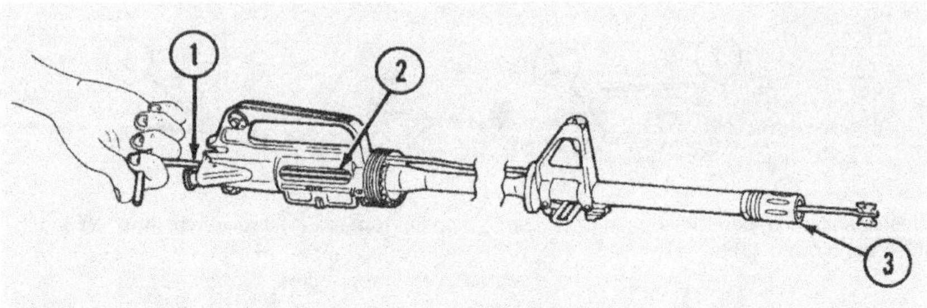

4 Take off bore brush and install swab holder and a swab (item 5, app D). Apply CLP (item 1, app D) to the swab. Run rod (1) through chamber (2) and flash suppressor (3) and back.

5 To remove carbon, let CLP (item 1, app D) set. Then wipe off clean. A bore brush may be used with care to loosen heavy buildup of carbon.

6 Wipe dry by running rod (1) with swab holder with clean swabs (item 5, app D) through chamber (2) and flash suppressor (3).

3-6. MAINTENANCE OF UPPER RECEIVER AND BARREL ASSEMBLY (CONT).

INSPECTION

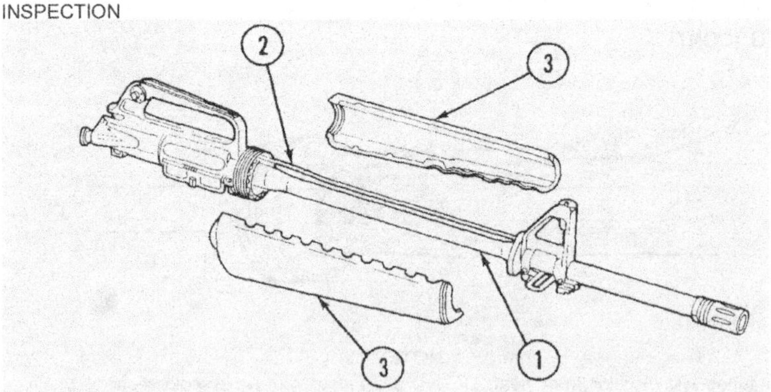

Inspect barrel (l), gas tube (2), and hand guards (3) for cracks, bends, or breaks. If you think apart is bad, notify organizational maintenance.

LUBRICATION

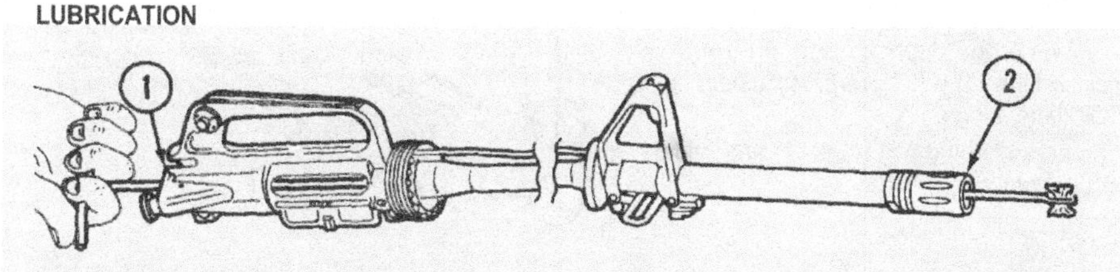

1 Lightly lubricate bore and chamber, outer surface of barrel and front sight, and surfaces under hand guards.

2 Start at receiver (1) and go right through the flash suppressor (2). Don't reverse directions in bore.

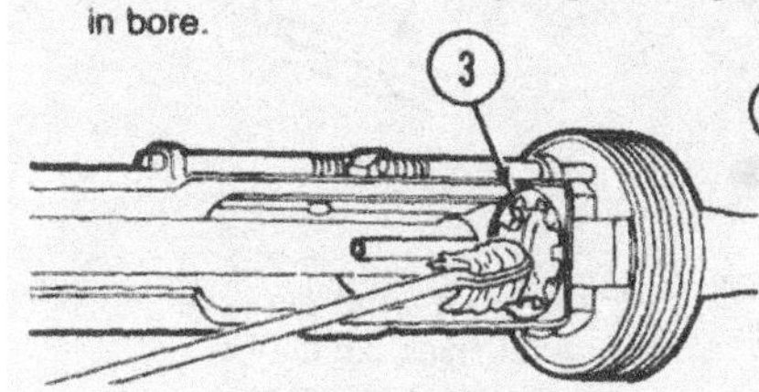

3 Lubricate locking lugs (3).

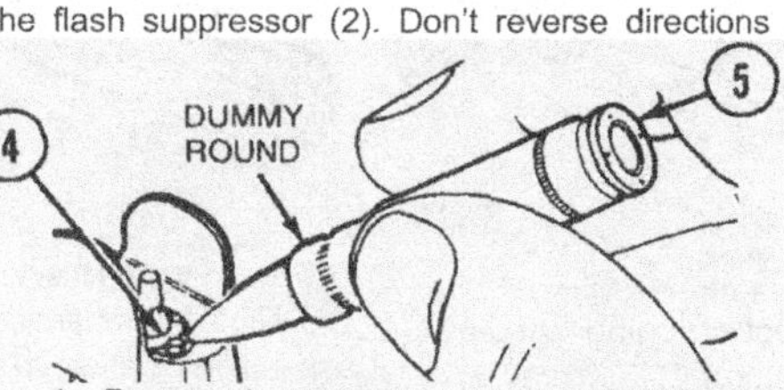

4 Depress front sight detent (4) several times to work CLP (item 1, app D) into the spring using a round (5).

3-7. MAINTENANCE OF CHARGING HANDLE ASSEMBLY AND BOLT CARRIER ASSEMBLY.

CLEANING

CAUTION

Do not use firing pin to clean inner surfaces of bolt or bolt carrier assembly,

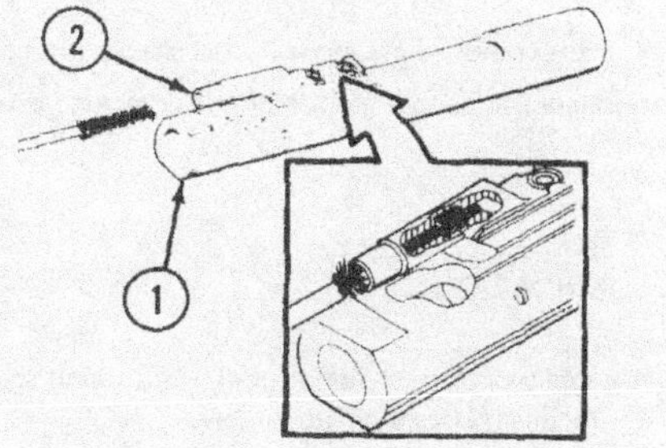

1 Clean all parts and outer surfaces of the bolt carrier assembly (1) with a swab (item 5, app D) saturated with CLP (item 1, app D).

2 Clean the bolt carrier key (2) with a worn bore brush dipped in CLP (item 1, app D). Dry with a pipe cleaner(item 2, app D) and swab (Item 5, app D). Use a pipe cleaner (item 2, app D) to apply a light coating of CLP (item 1, app D) to carrier key (2).

3 Remove carbon deposits and dirt from locking lugs (3) with bore brush dipped in CLP (item 1, app D),

4 Clean areas behind bolt rings (4) and under lip of extractor (5),

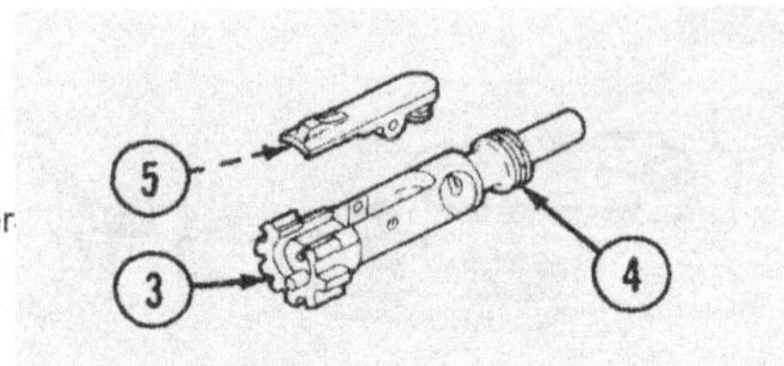

INSPECTION

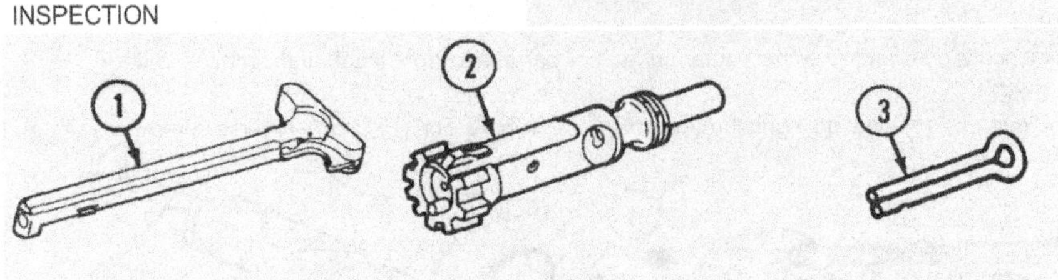

1 Inspect charging handle assembly (1) for cracks, bends, or breaks.

WARNING

DO NOT interchange bolt assemblies between rifles.

2 Inspect bolt assembly (2) for cracks or fractures, especially in the cam pin hole area.

3 Inspect firing pin retaining pin (3) for bends, breaks, or dents.

3-7. MAINTENANCE OF CHARGING HANDLE ASSEMBLY AND BOLT CARRIER ASSEMBLY (CONT).

INSPECTION (CONT)

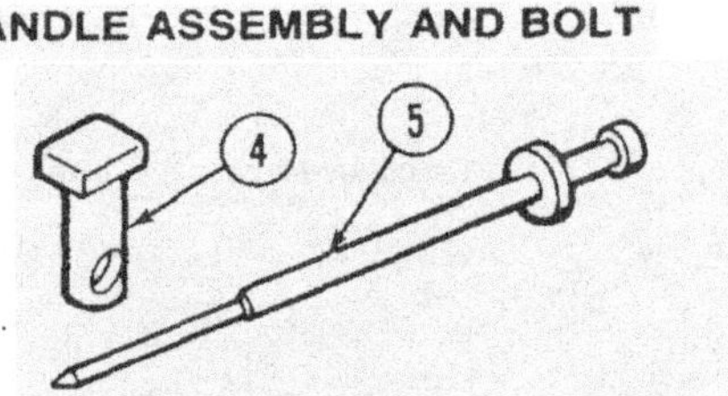

4 Inspect bolt cam pin (4) for cracks or chips.

5 Inspect firing pin (5) for bends, cracks, or blunted tip.

LUBRICATION

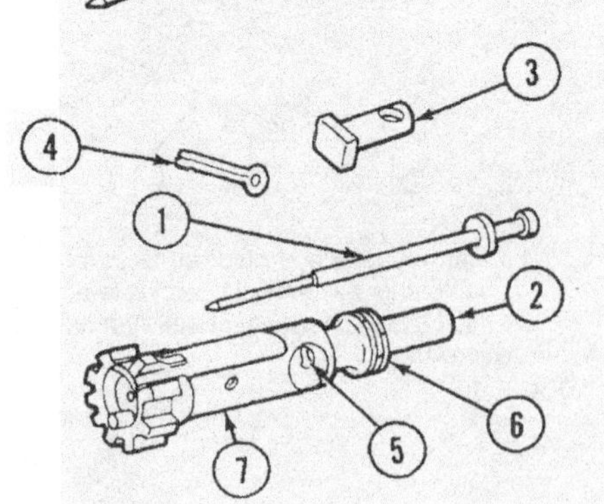

1 Lightly lubricate firing pin(l) and firing pin recess (2) in bolt assembly.

2 Generously lubricate outside of bolt cam pin (3) and firing pin retaining pin (4) with CLP (item 1, app D). Make certain to lubricate bolt assembly cam pin hole (5), bolt rings (6), and outside of the bolt assembly (7).

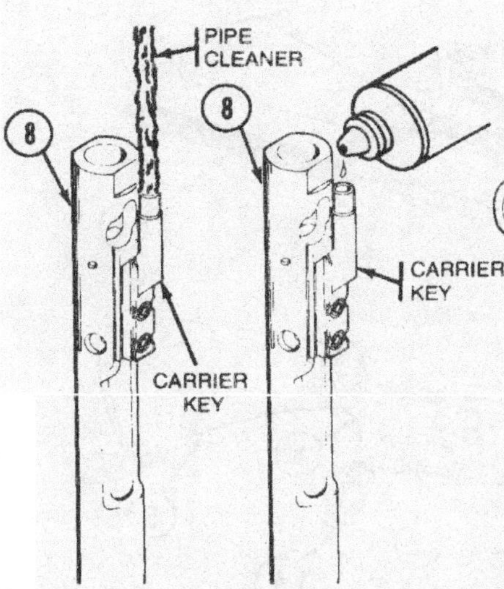

3 **Dry inside key of bolt carrier assembly (8). Place one drop of CLP (item 1, app D) inside key.**

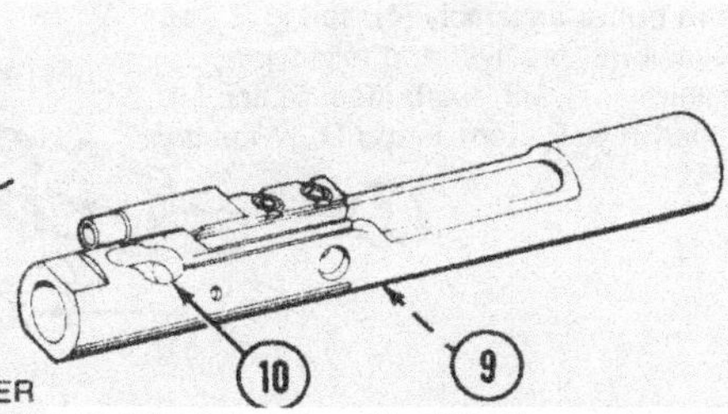

4 Lightly lubricate with CLP (item 1, app D) inner and outer surfaces of bolt carrier assembly (8). Generously lubricate slide (9) and cam pin area (1 O) of bolt carrier assembly (8).

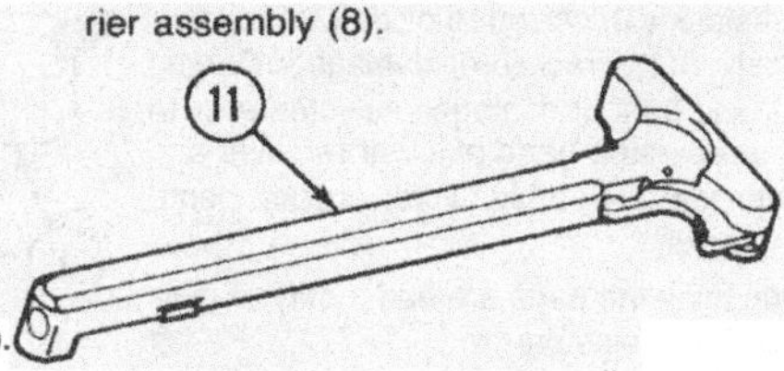

5 Lightly lubricate charging handle assembly (1 1).

3-8. MAINTENANCE OF LOWER RECEIVER AND EXTENSION ASSEMBLY.

CLEANING

CAUTION

Do not use steel/wire brush or any type of abrasive material to clean aluminum surfaces.

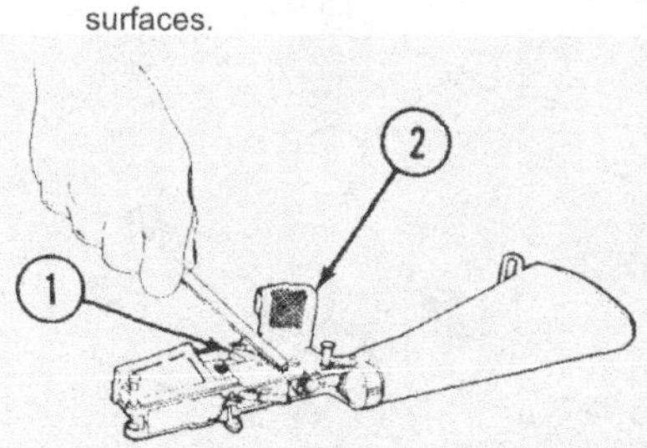

1 Wipe dirt from trigger (1) with a swab (item 5, app D).

and cleaning brush to clean powder fouling, corrosion, and dirt from outside parts of lower receiver and extension assembly (2)

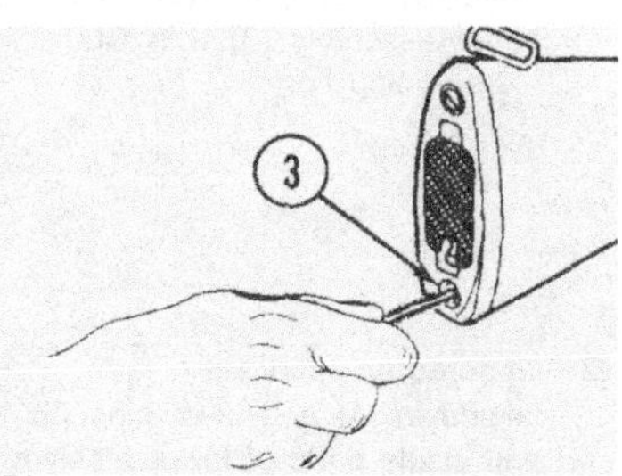

3 Use pipe cleaner (item 2, app D) to clean buttstock screw drain hole (3),

4 Clean buffer assembly (4), spring (5), and inside lower receiver and extension assembly (2) with swab (item 5, app D) dipped in CLP (item 1, app D). Wipe dry.

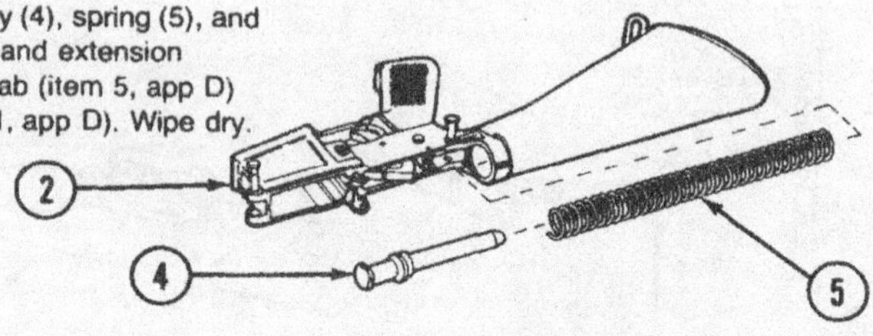

INSPECTION

1 Examine lower receiver and extension assembly (1) for broken or bent trigger (2), buttstock (3), corroded or deformed lower receiver (1), cracked or damaged rifle grip (4), and bent or damaged selector lever(5). Look at inside parts of lower receiver and extension assembly (1) for cracks, dents, or breaks.

2 If you think the parts are bad, notify organizational maintenance.

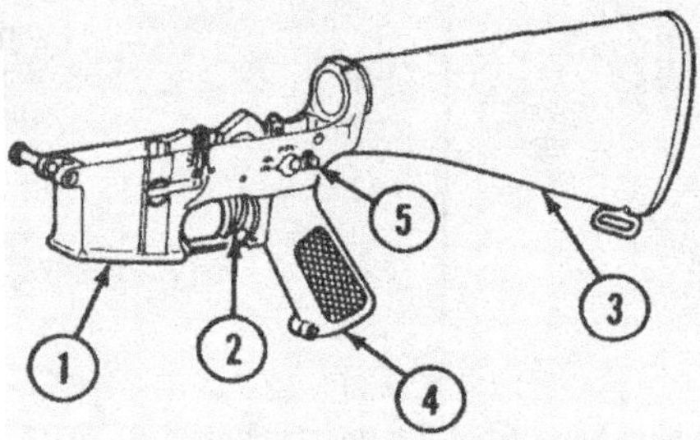

3-8. MAINTENANCE OF LOWER RECEIVER AND EXTENSION ASSEMBLY (CONT).

LUBRICATION

1 Lightly lubricate inside of lower receiver and extension assembly (1), spring (2), and buffer assembly (3) with CLP (item 1, app D).

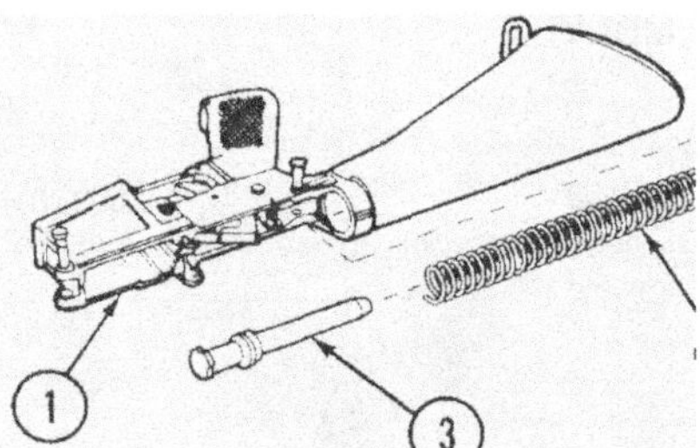

2 Generously lubricate takedown (4) and pivot pins (5) and inside parts of lower receiver and extension assembly (1) with CLP (item 1, app D).

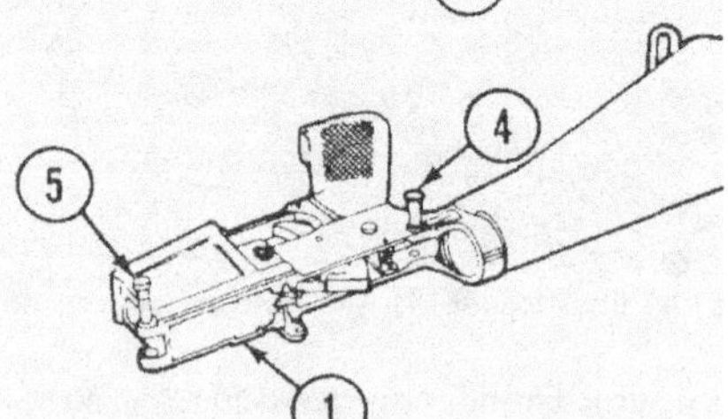

. REASSEMBLY OF M16/M16A1 RIFLE.

LOWER RECEIVER AND EXTENSION ASSEMBLY

Insert spring (1) and buffer (2).

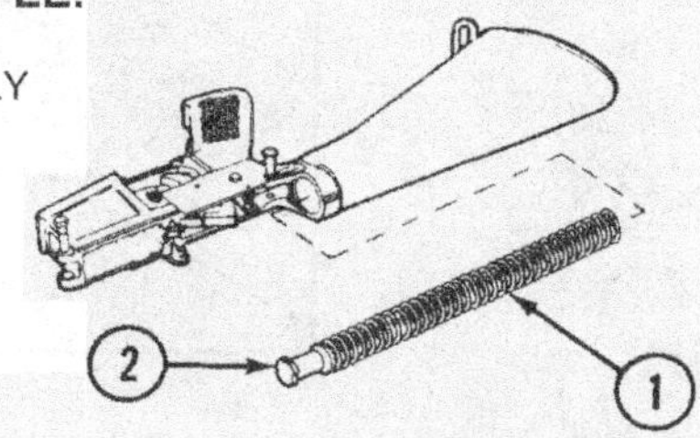

BOLT CARRIER ASSEMBLY AND CHARGING HANDLE ASSEMBLY

NOTE

New extractor has silicone insert with spring. Be sure not to lose it.

1 If the spring comes loose, seat the large end of spring in the extractor,

2 Insert extractor (1) and spring assembly (2) into bolt.

3 Push extractor (1) and spring assembly (2) down. Aline hole (3) with hole in bolt and insert extractor pin (4).

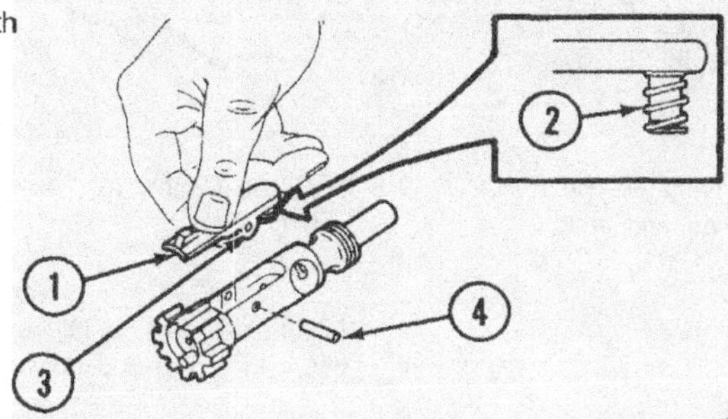

3-9. REASSEMBLY OF M16/M16A1 RIFLE (CONT).

BOLT CARRIER ASSEMBLY AND CHARGING HANDLE ASSEMBLY (CONT)

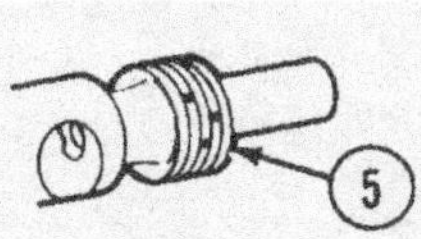

4 Stagger gaps in bolt rings (5) to stop gas loss.

WARNING

Don't switch bolt assemblies between rifles.

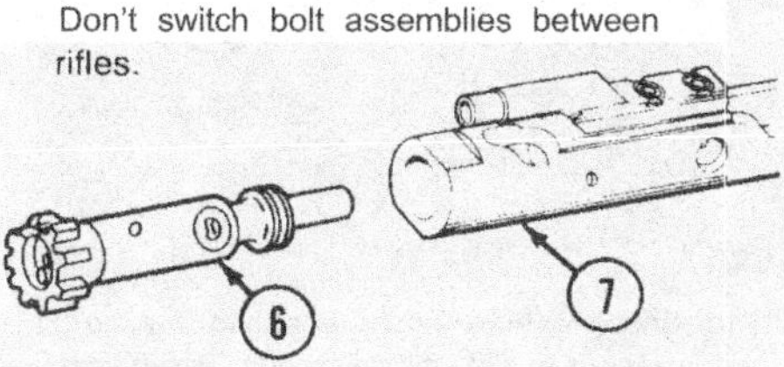

5 Slide bolt assembly (6) into bolt carrier assembly (7).

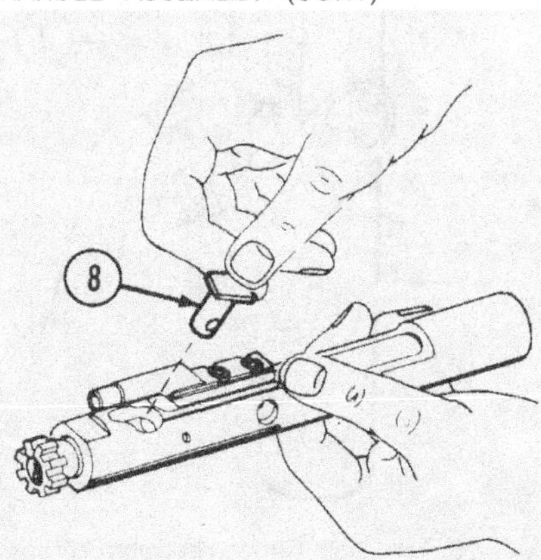

6 Insert bolt cam pin (8) and give it a 1/4 turn.

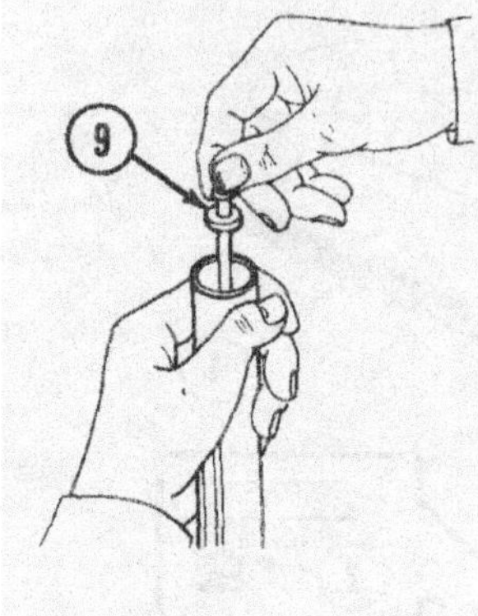

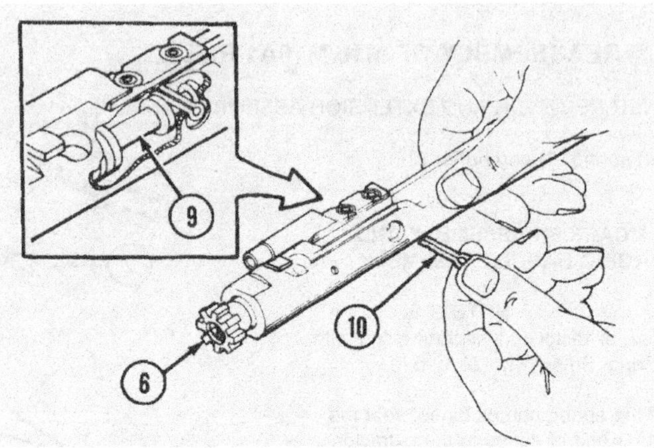

7 Drop firing pin (9) in opening and seat.

8 Pull boltassembly (6) back and replace firing pin retaining pin (10).

NOTE

Firing pin should not fall out when bolt carrier assembly is turned upside down.

3-9. REASSEMBLY OF M16/M16A1 RIFLE (CONT).

BOLT CARRIER ASSEMBLY AND CHARGING HANDLE ASSEMBLY (CONT)

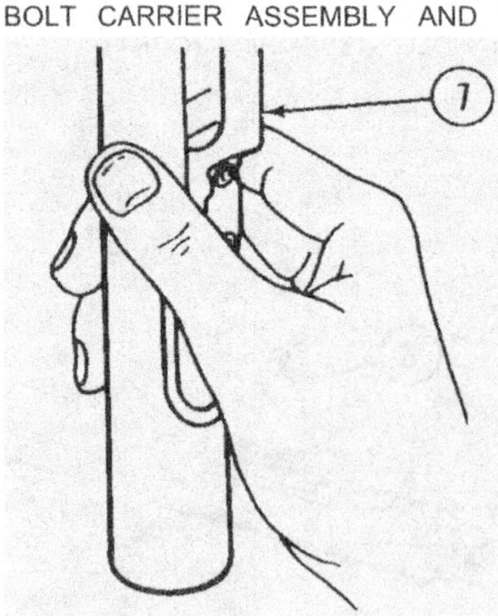

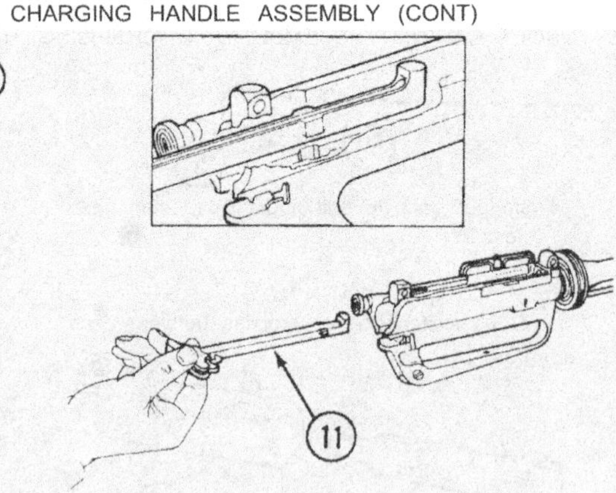

9 Turn bolt carrier assembly (7) over and try to shake out firing pin.

10 Engage, then push, charging handle assembly (11) part way into upper receiver.

NOTE

Be sure bolt assembly is extended from bolt carrier.

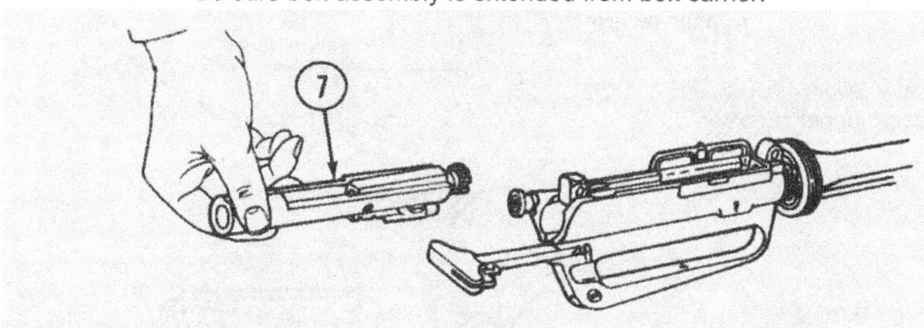

11 **Slide bolt carrier assembly (7) into upper receiver.**

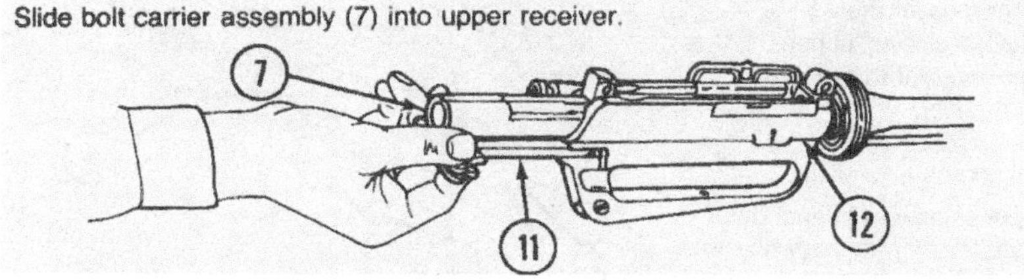

12 Push charging handle assembly (11) and bolt carrier assembly (7) together into upper receiver (12).

3-9. REASSEMBLY OF M16/M16A1 RIFLE (CONT).

JOINING UPPER AND LOWER RECEIVERS

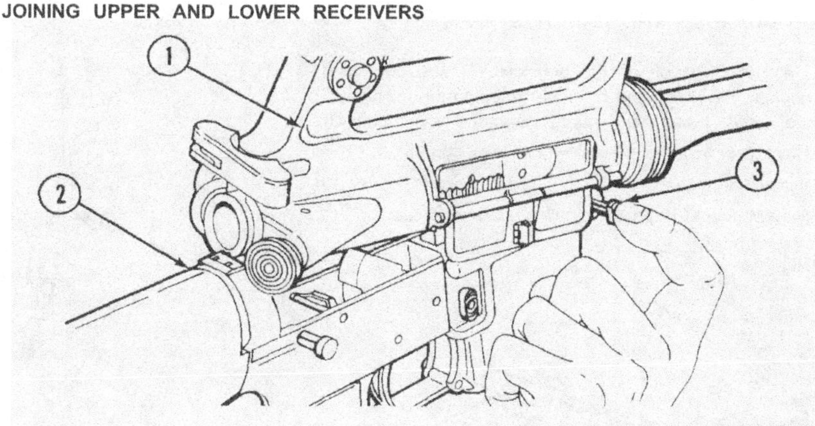

1 Join upper receiver (1) and lower receiver (2).

2 Aline the pivot pin holes and push pivot pin (3) in.

CAUTION

Selector lever must be on SAFE or SEMI before closing upper receiver.

3 Place selector lever (4) on SAFE or SEMI before closing upper receiver.

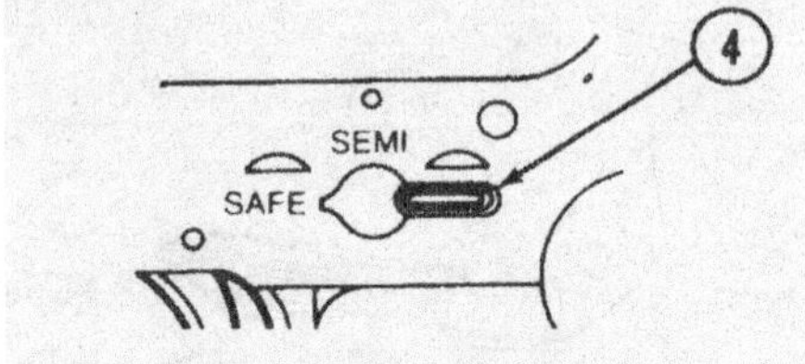

CAUTION

Ejection port cover must be closed before closing upper and lower receiver to prevent damage to cover.

4 Close ejection port cover (5).

5 Close upper receiver (1) and lower receiver (2). Push in takedown pin (6).

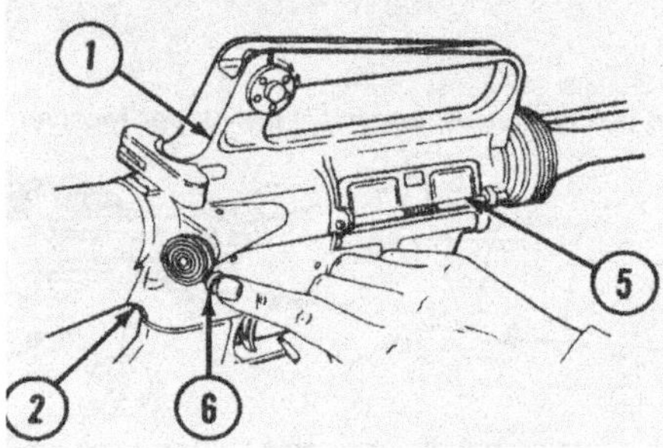

3-9. REASSEMBLY OF M16/M16A1 RIFLE (CONT)

HAND GUARDS - THE "BUDDY SYSTEM"

1 Place the weapon on the buttstock (1) with one hand gripping the stock and the other gripping the lower end of the barrel. Insert hand guard into hand guard cap (2).

2 Have your buddy press down on slip ring (3).

3 Install hand guard (4).

4 Repeat these steps to install other hand guard.

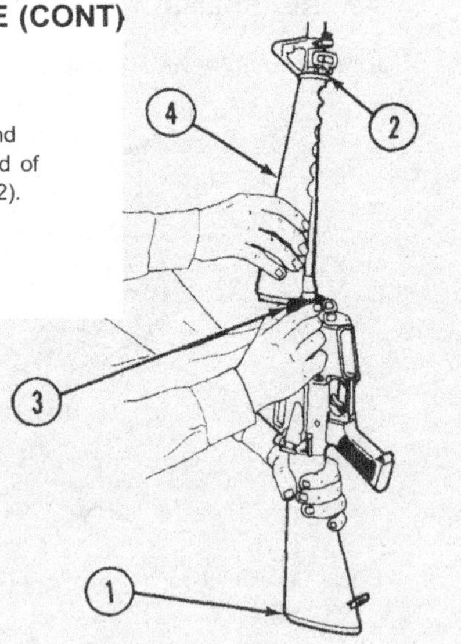

SLING

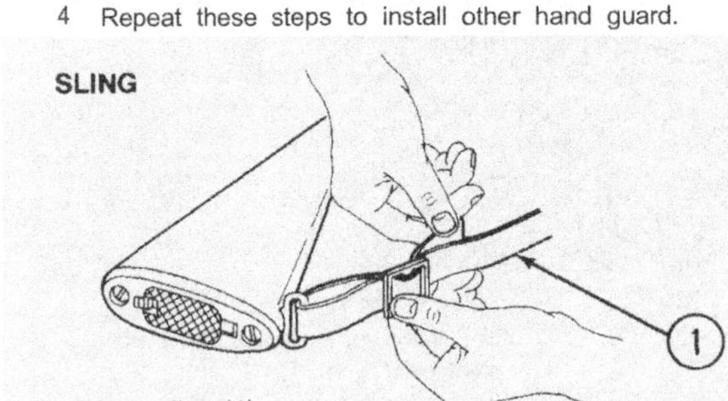

Attach sling (1).

3-10. FUNCTIONAL CHECK.

WARNING

To avoid accidental firing, be sure cartridge magazine is removed and chamber is clear (p 2-34).

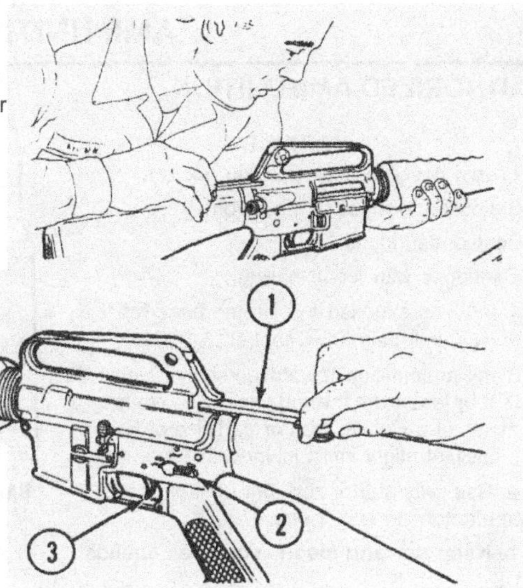

1 **Pull charging handle assembly (1) to rear and release. Place selector lever (2) on SAFE. Squeeze trigger (3). Hammer should not fall.**

3-10. FUNCTIONAL CHECK (CONT).

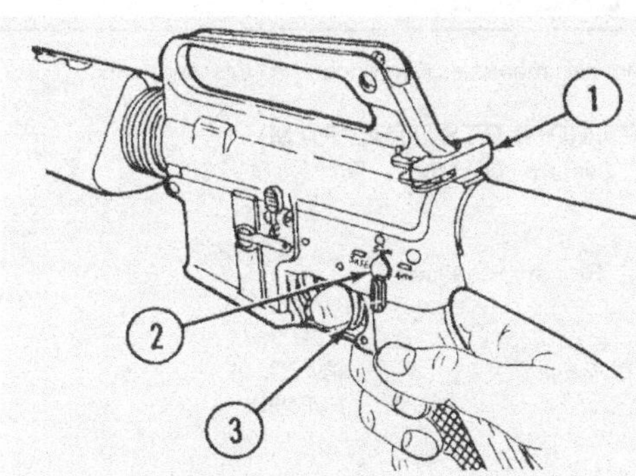

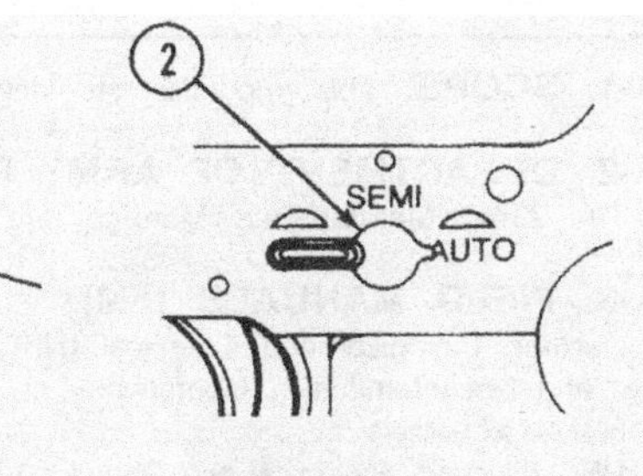

2 Place selector lever (2) on SEMI. Squeeze trigger (3); hammer should fall. Hold trigger to the rear. Pull charging handle assembly (1) to rear and release. Release trigger (3). You should hear a click as you release the trigger. Squeeze again; hammer should fall.

3 Place selector lever (2) on AUTO. Pull charging handle assembly (1) to rear and release, Squeeze the trigger; hammer should fall. Hold trigger to the rear and cock the rifle. Release the pressure on the trigger and squeeze it to the rear again. The hammer should not fall because it should have fallen when the bolt was allowed to move forward during the cocking sequence.

CHAPTER 4
AMMUNITION

AUTHORIZED AMMUNITION.

WARNING

To avoid possible explosion, do not fire:

- Seriously corroded ammunition
- Dented cartridges
- Cartridges with loose bullets
- Cartridges exposed to extreme heat 135°F (57°C) until they have cooled
- Blank ammunition toward personnel within 20 feet or less from the muzzle, because fragments of a closure wad or particles of unburnt propellant might inflict injury within that range.

a. Use only authorized ammunition that is manufacture to US specs.

b. **Keep dry and clean.** Your life depends on it!

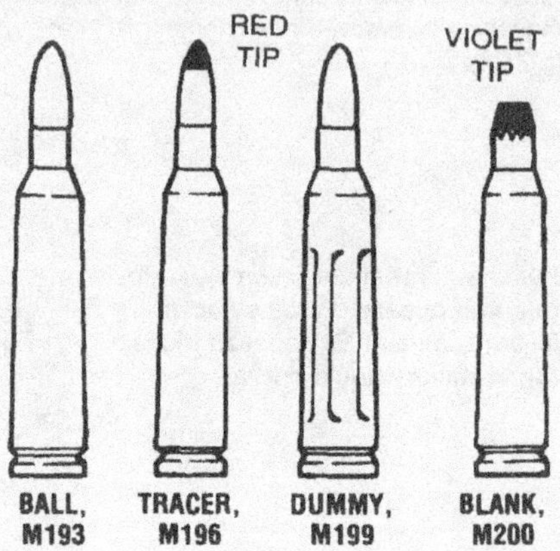

APPENDIX A
REFERENCES

A-1. SCOPE. This appendix lists techmical manuals referenced in this manual.

A-2. DEPARTMENT OF ARMY PAMPHLETS (DA PAM).

The Army Maintenance Management System (TAMMS) DA PAM 738-750

A-3. FIELD MANUALS (FM).

Nuclear, Biological and Chemical (NBC) Reconnaissance and Decontamination Operations . . . . FM 3-87

First Aid for Soldiers . . . . FM 21-11

NBC (Nuclear, Biological and Chemical) Defense . . .. FM 21-40

M16A1 Rifle and Rifle Marksmanship . . . .FM 23-9

A-4. TECHNICAL MANUALS (TM).

Chemical, Biological and Radogical (CBR) Decontamination .. . . TM 3-220

APPENDIX B
COMPONENTS OF END ITEM AND BASIC ISSUE ITEMS LISTS

Section I. INTRODUCTION

B-1. SCOPE. This appendix lists components of end item and basic issue items for the rifle to help you inventory items required for safe and efficient operation.

B-2. GENERAL. The Components of End Item and Basic Issue Items Lists are divided into the following sections:

a. *Section //. Components of End Item.* This listing is for informational purposes only, and is not authority to requisition replacements. These items are part of the end item, but are removed and separately packaged for transportation or shipment. As part of the end item, these items must be with the end item whenever it is issued or transferred between property accounts. Illustrations are furnished to assist you in identifying the items.

b. *Sectlon III. Basic Issue Items.* These are the minimum essential items required to place the rifle in operation, to operate it, and to perform emergency repairs. Although shipped separately packaged, BII must be with the rifle during operation and whenever it is transferred between property accounts. The illustrations will assist you with hard-to-identify items. This manual is your authority to request/requisition replacement BII, based on TOE/MTOE authorization of the end item.

B-3. EXPLANATION OF COLUMNS. The following provides an explanation of columns found in the tabular listings:

a. *Column (1) - Illustration Number (Illus Number).* This column indicates the number of the illustration in which the item is shown.

b. *Column (2) - National Stock Number.* Indicates the National stock number assigned to the item and will be used for requisitioning purposes.

c. *Column (3) - Description.* Indicates the Federal item name and, if required, a minimum description to identify and locate the item. The last line for each item indicates the FSCM (in parentheses) followed by the part number.

d. Column (4) - *Unit of Measure (U/M).* Indicates the measure used in performing the actual operational/maintenance function. This measure is expressed by a two-character alphabetical abbreviation (e.g., ea, in., pr).

e. *Column (5) - Quantify required (Qty rqr).* Indicates the quantity of the item authorized to be used with/on the equipment.

Section II. COMPONENTS OF END ITEM

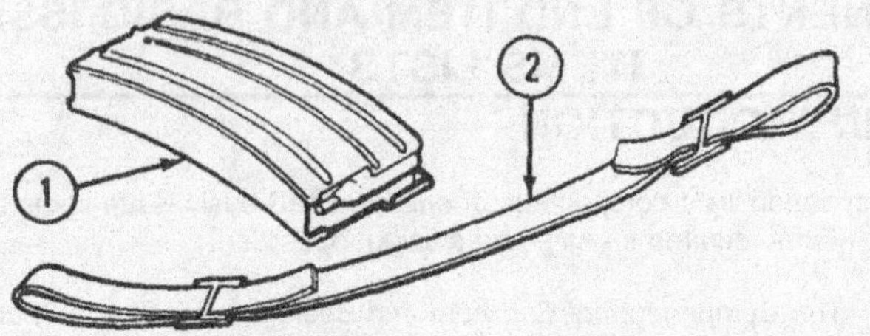

(1) Illus Number	(2) National Stock Number	(3) Description FSCM and Part Number	(3) Usable On Code	(4) U/M	(5) Qty rqr
1	1005-00-921-5004	MAGAZINE, CARTRIDGE: 30 round (19204) 8448670		EA	1
2	1005-00-167-4336	SLING, SMALL ARMS (19204) 8448770		EA	1

Section III. BASIC ISSUE ITEMS

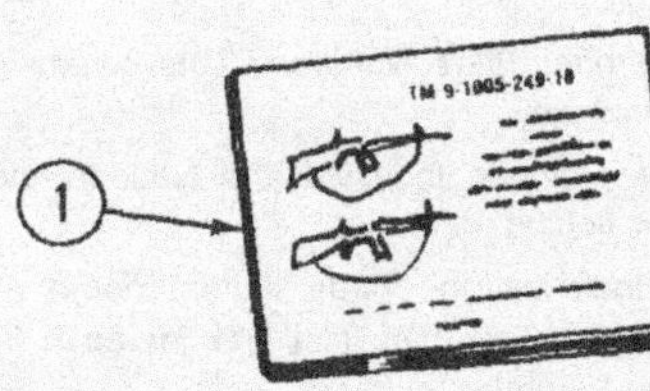

(1) Illus Number	(2) National Stock Number	(3) Description FSCM and Part Number	(3) Usable On Code	(4) U/M	(5) Qty rqr
1		TM 9-1005-249-10		EA	1

APPENDIX C
ADDITIONAL AUTHORIZATION LIST (AAL)

Section 1. INTRODUCTION

C-l. SCOPE. This appendix lists additional items you are authorized for the support of the rifle.

C-2. GENERAL. This list identifies items that do not have to accompany the rifle and that do not have to be turned in with it. These items are all authorized to you by CTA, MTOE, TDA, or JTA.

C-3. EXPLANATION OF LISTING. National stock numbers, descriptions, and quantities are provided to help you identify and request the additional items you require to support this equipment. The items are listed in alphabetical sequence by item name under the type document (i.e., CTA, MTOE, TDA, or JTA) which authorizes the item(s) to you.

Section II. ADDITIONAL AUTHORIZATION LIST (AAL)

(1) NATIONAL STOCK NUMBER	(2) DESCRIPTION FSCM AND PART NUMBER	USABLE ON CODE	(3) U/M	(4) QTY AUTH
	CTA AUTHORIZED ITEMS			
1005-00-193-8306	BAG, PROTECTIVE: for 30 round magazine (500 per bx) required 1 per magazine (1 9204) 8448464		EA	1
1005-00-118-6192	BLANK FIRING ATTACHMENT, M15A2: (For training only) (1 9204) 12002900		EA	1
1005-00-903-1296	BRUSH, CLEANING, SMALL ARMS: bore (1 9204) 11686340		EA	1
1005-00-999-1435	BRUSH, CLEANING, SMALL ARMS: chamber (1 9204) 8432358		EA	1

Section II. ADDITIONAL AUTHORIZATION LIST (AAL) (CONT)

(1) NATIONAL STOCK NUMBER	(2) DESCRIPTION FSCM AND PART NUMBER	USABLE ON CODE	(3) U/M	(4) QTY AUTH
1005-00-494-6602	BRUSH, CLEANING, SMALL ARMS: tooth (19204) 8448462		EA	1
5340-00-880-7666	CAP, PROTECTIVE, DUST [19204) 8445067		EA	1
1005-01-171-4778	CARTRIDGE CASE DEFLECTOR ASSEMBLY (19200) 9378328		EA	1
8465-00-781-9564	CASE, MAINTENANCE EQUIPMENT: for rifles without buttstock stowage (81349) MIL-C-43737 or		EA	1
1005-00-403-5804	CASE, SMALL ARMS: for rifles with buttstock stowage (19204) 8448751		EA	1

Section II. ADDITIONAL AUTHORIZATION LIST (AAL) (CONT)

NATIONAL STOCK NUMBER	(2) DESCRIPTION FSCM AND PART NUMBER	USABLE ON CODE	(3) U/M	(4) QTY AUTH
1005-01-113-0321	HANDLE SECTION, CLEANING ROD, SMALL ARMS (1 9204) 8436776		EA	1
1005-00 -406-1570	KIT, ADAPTER SLING (1 9204) 8448471		EA	1
1005-00-921-5004	MAGAZINE, CARTRIDGE: 30 round (1 9204) 8448670		EA	6

Section II. ADDITIONAL AUTHORIZATION LIST (AAL) (CONT)

(1) NATIONAL STOCK NUMBER	(2) DESCRIPTION FSCM AND PART NUMBER	USABLE ON CODE	(3) U/M	(4) QTY AUTH
1005-00-233-9031	PLATE, LOCKING: for riot control use, prevents selector from automatic fire (refer to organizational maintenance for installation and instructions on use) (19204) 8448676		EA	1
1005-00-050-6357	ROD SECTION, CLEANING ROD, SMALL ARMS (19204) 8436775		EA	3
1005-00-937-2250	SWAB HOLDER SECTION, CLEANING ROD, SMALL ARMS (19204) 11686327		EA	1

Section II. ADDITIONAL AUTHORIZATION LIST (AAL) (CONT)

(1) NATIONAL STOCK NUMBER	(2) DESCRIPTION FSCM AND PART NUMBER	USABLE ON CODE	(3) U/M	(4) QTY AUTH
	MTOE AUTHORIZED ITEMS			
1005-00-017-9701	BAYONET - KNIFE M7 W/SCABBARD (1 9204) 8427025		EA	1
1005-00-890-2609	BIPOD, RIFLE M3 W/CARRYING CASE (1 9204) 8445081		EA	1
1005-00-406-1570	TOP SLING ADAPTER KIT (1 9204) 8448471		EA	1

APPENDIX D
EXPENDABLE/DURABLE SUPPLIES AND MATERIALS LIST (EDSML)

Section I. INTRODUCTION

D-1. SCOPE. **This appendix lists expendable supplies and materials you will need to operate and maintain the M16/M16A1 rifle. This listing is for informational purposes only and is not authority to requisition the listed items. These items are authorized to you by CTA 50-970, Expendable/Durable Items (Except Medical, Class V, Repair Park, and Heraldic Items), or CTA 8-100, Army Medical Department Expendable/Durable Items.**

D-2. EXPLANATION OF COLUMNS.

a. Column (1) - Iterm Number. This number is assigned to the entry in the listing and is referenced in the narrative instructions to identify the material (e.g., "Use protective cover (item 3, app D)").

D-2. EXPLANATION OF COLUMNS (CONT).

b. *Column (2) - Level.* This column identifies the lowest level of maintenance that requires the listed item.

C - Operator/Crew

c. *Column (3) - National Stock Number.* This is the National stock number assigned to the item; use it to request or requisition the item.

d. *Column (4) - Description.* Indicates the Federal item name and, if required, a description to identify the item. The last line for each item indicates the Federal Supply Code for Manufacturer (FSCM) in parentheses followed by the part number.

e. *Column (5) - Unit of Measure (U/M).* Indicates the measure used in performing the actual maintenance function. This measure is expressed by a two-character alphabetical abbreviation (e.g., ea, in., pr). If the unit of measure differs from the unit of issue, requisition the lowest unit of issue that will satisfy your requirements.

Section II. EXPENDABLE SUPPLIES AND MATERIALS LIST (ESML)

(1) ITEM NO.	(2) LEVEL	(3) NATIONAL STOCK NUMBER	(4) DESCRIPTION	(5) U/M
1	C	9150-01-102-1473	CLEANER, LUBRICANT AND PRESERVATIVE: 1/2-oz bottle (81349) MIL-L-63460	OZ
2	C	9920-00-292-9946	CLEANER, TOBACCO PIPE: cotton tuft, wire core (32 per pk) (89855) DILLSPIPECLEANER	EA

EXPENDABLE SUPPLIES AND MATERIALS LIST (ESML) (CONT)

(1) ITEM NO.	(2) LEVEL	(3) NATIONAL STOCK NUMBER	(4) DESCRIPTION	(5) U/M
3	c	1005-00-809-2190	COVER, PROTECTIVE (1 9204) 8448213	EA
4	c	7920-00-205-1711	RAG, WIPING: 50 lb bdl (58536) A-A-531	LB
5	c	1005-00-912-4248	SWAB, SMALL ARMS CLEANING: cotton 1 pkg (1000 per package) (1 9204) 11686408	EA

APPENDIX E
STOWAGE GUIDE

Section I. INTRODUCTION

E-1. SCOPE. This appendix shall be included only when specified by the procuring activity.

E-2. GENERAL. This guide shall detail the physical installation/stowage location of all applicable AAL and expendable supply items required to be carried with the equipment.

Section II. GUIDE FOR STOWING ACCESSORIES IN SMALL ARMS ACCESSORIES CASE OR IN BUTTSTOCK

NOTE

The diagram is just a guide. It is not mandatory to assemble the contents according to the diagram.

MAINTENANCE EQUIPMENT CASE: SMALL ARMS ACCESSORIES, PACKED WITH CLEANING EQUIPMENT AND SUPPLIES

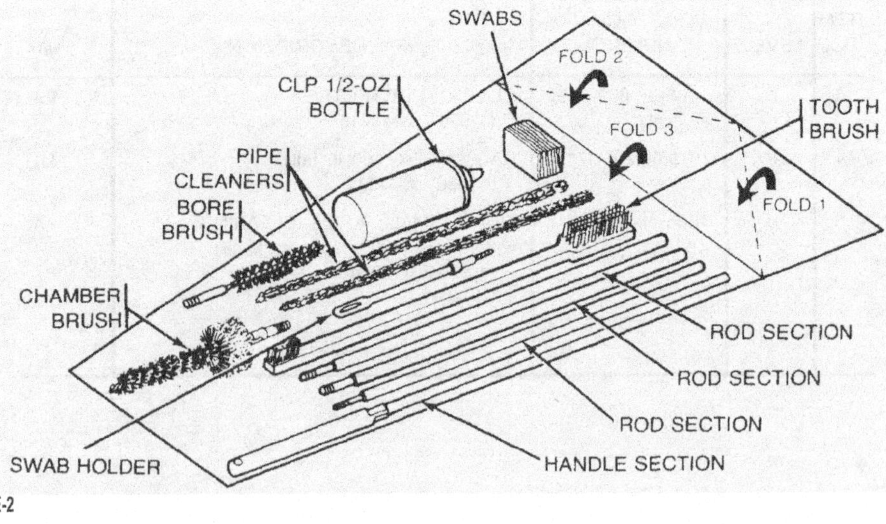

NOTE

See illustration below on how to open buttstock on rifles with buttstock storage.

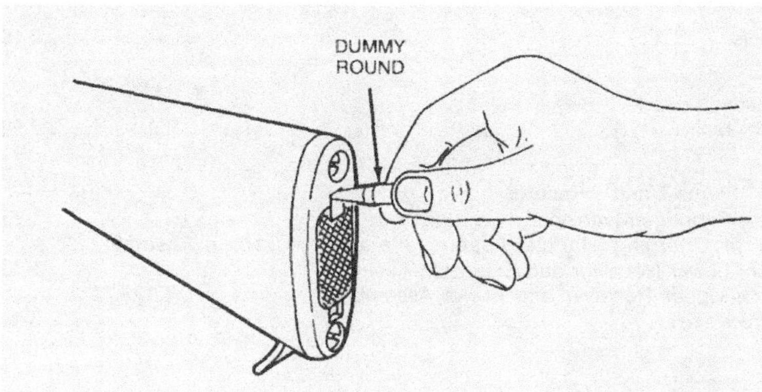

ALPHABETICAL INDEX

Page

G

I

L

M

O

R

s

T

BY order of the Secretary of the Army:

JOHN A. WICKHAM, JR.
General, United States Amy
Chief of Staff

official:

DONALD J. DELANDRO
Brigadier General, united States Amy
The Adjutant General

Distribution:

TO be distributed in accordance with DA Form 12-40, Operator and Crew Maintenance requirements for Rifle, 5.56MM, M16, M16A1.

☆ U.S. G.P.O.: 1994 300-421/00085

Made in the USA
Monee, IL
13 September 2026

62976953R00046